The Greening of Faith

God, the Environment, and the Good Life

THE GREENING OF FAITH

God, the Environment, and the Good Life

Edited by
John E. Carroll
Paul Brockelman
Mary Westfall

Foreword by
Bill McKibben

University of New Hampshire

Published by University Press of New England
Hanover and London

University of New Hampshire
Published by University Press of New England, Hanover, NH 03755
© 1997 by The Trustees of the University of New Hampshire; Foreword © 1997 by Bill
McKibben
Printed in the United States of America   5 4 3 2 1
CIP data appear at the end of the book

Contents

Foreword | BILL MCKIBBEN

For a very long time, people we have thought of as spiritual leaders have recommended similar things—simplicity, a pleasure in service, contact with one's fellows and with the natural world, humility. From Buddha and Jesus through St. Francis and Thoreau, straight through to Gandhi, we have tended to idealize these figures in theory, and in practice to ignore them as cranks.

One of the many features that makes the present moment so interesting is the emergence of another caste of shamans with similar conclusions. This time, they're atmospheric chemists, physicists, and the like. And they are advising simplicity, community—perhaps not in so many words, but that is the clear implication of their data about phenomena like global warming. Only a new idea of who we are would allow for the changes necessary to reverse the damage—would allow us to replace cars with buses, for instance. In our hyperindividualism, our hypermaterialism, we have simply become too large for the planet.

At heart, this question of scale is a religious question. Who is at the center of affairs? We have behaved for quite a long time as if we were, and it is this assumption that drives most of the environmental dilemmas we face. (It may cause many of the spiritual emptinesses we confront as well.) Now we need somehow to rediscover an older formula: that the tribe or nature or God, or most likely some amalgam of the three, is at the center of things. In such a world it becomes possible to imagine certain limits, to imagine fulfilling our unique ability as a species to limit ourselves.

Theologians and religious institutions will be of enormous help in this quest, as this volume makes clear. No other institutions and virtually no other intellectuals are even potentially immune to the infection that several writers herein describe as "growthism." Business believes it must grow; educators equip us to aid that growth; politicians see their role as speeding up growth ("it's the economy, stupid"). Only churches, synagogues, mosques (and the campfires of those who do their worshipping outdoors) acknowledge the possibility of some other goal.

In a way, as many of these essays hint, the moment offers enormous opportunity for people of faith. The path of scientific and economic idolatry seems to have turned into a box canyon; there is new opportunity to lay out more profound trails into the future. We can revivify our theological and institutional life along the lines suggested in this book, jettisoning many of the sterile formulations ill-suited for this adventure, telling new stories and very ancient ones.

But it is also a time of great peril, of course. For if present trends hold—if, inside a generation, people influence or dominate all the natural processes around them—then any attempt even to talk about God will become fraught with trouble. Humans have always sensed the divine in sunshine, in wind (*ruach*, the first wind), in rain and storm, in leaf and petal, in talon and tooth and tail, in caw and snarl and purr. If all those things become mere categories of human enterprise—of our alterations of climate, of our willingness to switch and shuffle genes—then we are so large that bowing down is less likely. We become so lonely, so radically lonely, so set apart from God and from creation.

Though the hour is extremely late and though much environmental damage is already guaranteed, it is not yet impossible to imagine us learning from this passage a new humility. It is possible that not only could we read this new/old theology but could begin swiftly to act on it and in that praxis discover its real meanings and joys. It is possible. But only if we start.

The Greening of Faith

God, the Environment, and the Good Life

Introduction
Getting Our Bearings

It used to be, before the days of electronic depth-sounding devices, that ships had to carefully "feel" their way over unfamiliar and shallow waters lest they go aground to their possible ruin. They did this by casting forward from the boat a carefully measured and weighted line that, when it reached bottom, would indicate the depth of water at just that moment and just that spot. It is not difficult to imagine such ships "sounding" their way home in a dense fog, slowly and softly gliding to their safe berth in their home port. The voice of a crewman rings out from the bow for the captain aft of him to hear: "mark 10," "mark 8½," "mark 7," and so on. Thus the ship inched forward, avoiding shallows and thus avoiding the sort of extreme peril and even disaster that awaited those too hurried, too ignorant, too unconcerned, or even too self-centered to look out for their own welfare and that of the whole ship.

It may be that we are at present facing a similar situation ecologically, for it would seem that our modern world is precipitously moving over increasingly shallow and dangerous waters, in which we risk our lives out of ignorance, self-absorption, or just plain lack of concern for anything larger than our own momentary cares and concerns. The analogy breaks down, of course, precisely because there is more at stake here than just we humans and our "civilization" (the crew and the ship). Life itself, at least as we have known it, seems at risk and already rather far along on the route to annihilation.[1]

Still, like those ships and their crews, we need some way not only to measure our dangers but to care enough to do something about them instead of—in our haste to make home—unconsciously sailing full-rigged and hell-bent for the rocks that seem to lie all about us. Science, of course, gives us the tools for sounding and thus measuring the depths over which we sail. But where shall we find the captain or crew who care, who hear the soundings and care enough to help us turn away from careless calamity? Where are we to find people who love life in all its forms enough not only to sound the warnings but also to heed them and to turn the ship toward safer waters? Where might we find those with the sensitivity toward nature in all its miraculous forms, a reverence and concern for all of creation deep and broad enough to overcome its casual destruction for abysmally self-centered and limited human concerns? Where else, indeed, might we find such people but in our churches, synagogues, mosques, and ashrams where creation is thought to be holy?

In fact, those spiritual traditions seem to be awakening to just such a reverential awareness for all of creation. So serious is the environmental situation thought to be that a number of national and international religious institutions and organizations such as the World Council of Churches, the U.S. Catholic Conference, the United Church of Christ, the Interfaith Council of the United Nations, and many others from most of the world's major religious traditions have made coming to terms with the environmental situation we now face a top priority. As one international and interfaith group of religious leaders put it recently,

We believe the environmental crisis is intrinsically religious. All faith traditions and teachings firmly instruct us to revere and care for the natural world. Yet sacred creation is being violated and is in ultimate jeopardy as a result of longstanding human behavior. A religious response is essential to reverse such longstanding patterns of neglect and exploitation.[2]

But changes in our religious attitudes toward nature may be necessary for another reason. Without the passion and discipline of religious life, it may be impossible actually to alter our behavior toward the environment. As economist Herman Daly and theologian John Cobb put it in their recent book, *For the Common Good,*

. . . a sustained willingness to change depends on a love of the earth that human beings once felt strongly, but that has been thinned and demeaned as the land was commodified. . . . there is a religious depth in myriads of people that can find expression in lives lived appropriately to reality. That depth must be touched and tapped. . . . If that is done, there is hope. . . . Our point is that the changes that are now needed in society are at a level that stirs religious passions. The debate will be a religious one whether that is made explicit or not.[3]

Such a shift in consciousness and attitude is, in fact, taking place in what has been variously called ecotheology, deep ecology, ecofeminism, or just plain environmental theology. This startling new development is taking place across various religious traditions and promises to be one of the most significant paradigm shifts in theology in this century. It entails a radical rethinking and reorientation of religious perspective and behavior in the light of the environmental deterioration we are currently witnessing. The authors of the essays that follow are clearly part of this shift in consciousness.

What our present situation calls for, then, is spiritual and moral renewal and reform, a new way of orienting ourselves and the way we live consistent with our fundamental visions of life and within the frame of nature as a whole. In these closing years of the twentieth century, we are witnessing a dramatic turn toward a deepened interest in the interrelationship of ecology and religion.

We are witnessing not only a strong turn within the environmental movement as a whole toward spiritual values and perspectives but also, in a related way, a heightened interest in the practical analysis and application of ethical values to stubborn environmental issues in energy and agriculture, the utilization of natural resources, and in general the sprawling encroachment of human development on the natural world. Within the so-called deep ecology movement, for example, there is a demand for a moral approach to nature that goes beyond picturing nature as a mere utility for human production and use to viewing it as intrinsically valuable in and of itself. In outlining the basic platform of deep ecology, the Norwegian philosopher Arne Naess put it this way: "The well being and flourishing of human and nonhuman life on earth have value in themselves (synonyms: intrinsic value, inherent value). These values are independent of the usefulness of the nonhuman world for human purposes."[4] Furthermore, there is a growing recognition that the various sciences, however indispensable and valuable they may be in dealing with human despoliation of the earth, are by themselves incomplete in helping us to behave and live on the earth in a more appropriate and sustainable manner.

As conventional wisdom has it, the environmental question is fundamentally a question of science and technology. A large segment of such conventional wisdom takes the matter to be an economic question. Another conventional approach has been to assume the matter to be political in nature, to be a question of politics, public policy, political science, law, or perhaps diplomacy or foreign affairs. Each of these types of conventional wisdom has assumed that the answer to the ecological or environmental question, the natural resources or energy question, the agriculture question, is to be found, therefore, in the scientific or technological fix, in economic management or manipulation, in political compromise or negotiation. And while each of these most certainly has a role to play and can make a contribution, we know that each of them and all of them together have fundamentally failed to stem the tide of environmental catastrophe.

Increasingly, there is a growing recognition that the value dimension must be an important element in any solution. Environmental issues demand both spiritual and ethical consideration and reflection. To paraphrase Kant, we might say that from an ecological point of view science without morality is disoriented; morality without science is empty. We need to get our bearings again: ecologically, ethically, and spiritually.

We ought to point out here that turning to ethical and spiritual reflection and value is neither an act of despair at the limitations of our natural and policy sciences nor a glissade into comforting illusion and self-delusion. This is not a kind of childish fantasy or wish fulfillment, a form of magical thinking in which "we" need do nothing because "God" will do it for us. The turn to value is not a sort of last resort because science, economics, and politics have failed us and in our fear we must withdraw into delusion. An ecology of despair and anxiety will not serve us well in the long run, even if it is capable of bringing about some short-term alleviation of the problem.

On the contrary, in the long run we will be safer and better served if instead of ecological anxiety we advocate a positive position of moral responsibility founded on a spiritual sense of our role and place within the deeper and encompassing reality of nature. We need an ecology of wonder and enchantment, a spiritual awareness of the intrinsic value or epiphany that nature manifests, and the proper sense of gratitude, humility, and awe that goes with it. There certainly is no lack of highly competent and credible people, scientists and others, who can rattle off for us the terrible litany of destruction already wrought upon our planetary ecosystem and ourselves. But what seems to be called for now is equally credible individuals who can help us to become aware of this sense of reverential wonder and to focus through it on the incredible beauty and complexity of creation.

The essays that follow are attempting just that. They are trying to bring us spiritually to such a strong sense of wonder at the incredible beauty and complexity of the ecosystem, of the planet as a whole. They remind us of how very little we really know, of how minuscule our knowledge is compared to what there is to know, and of the frontiers that remain undiscovered and even unexplored. They seek to leave us with a sense of enchantment so profound that we transcend much of our ability to do damage, to harm the fabric, the pattern, the functioning of the interdependent whole to which we belong.

Those who witness to the damage, and that alone, may do a disservice insofar as they may numb us into inaction. But those who help us to explore the magnificent and interdependent systems in which we are situated may help us to become enchanted and thereby learn our place within the whole. In the long run, that sympathetic spiritual opening to reality may do us, our successors, and our planetary ecosystem a much higher service. Above all, they may help us to avoid succumbing to the numbing death and decay toward which the ecology of anxiety inevitably leads.

A central thread, then, uniting the authors and essays that make up this volume is the belief that the environment is not only a spiritual issue but the spiritual issue of our time. Their words and vision reinforce this theme. In the words of Dean James P. Morton of the Episcopal Cathedral of Saint John the Divine, "the environment is not just another issue but an inescapable challenge to what it means to be religious." It is to those who seek to understand what the challenge might mean in this, the final decade of the twentieth century that these voices are directed. This is by no means an exhaustive review of all major world religions and their critical offerings on "The Greening of Faith." The intended audience for this book is just those religious traditions that have most shaped the North American religious ethos in the past or seem situated to do so now. Thus, we have made no attempt to include all of the world's major traditions—an impossible task in any case. However, by focusing this work on those prevailing religious attitudes and practices that may bear the greatest responsibility for our current environmental crisis as well as our hope to overcome it and by making available newer emerging voices in our culture that challenge those attitudes, we hope the reader may find both useful critique and compelling religious vision to fit our particular time and place.

As the authors of these essays point out in their various ways, the dream of a healthy spiritual and ecological life and the dream of a more modest and appropriate economic life to go with it—all these are important but dreams nonetheless. That is to say that we have not yet achieved them; they are not realized but lie out ahead of us as vaguely envisaged half-realities. Of course, without them it would seem impossible to improve our future, for what they are, after all, are simply our human yearnings for a better and more fulfilling life. Like the great Mosaic dream of the Promised Land—a dream that has nurtured and shaped the lives of numberless Jews, Christians, and Muslims—these dreams beckon us to go beyond ourselves in order to make them real.

It is especially important just now to get our spiritual bearings, to get in touch with what fundamentally matters to us in living, for without that it may be difficult to find our way out of the ecological miasma in which we find ourselves. To do that—to get our spiritual bearings—is precisely to discover a thread connecting us to the rest of nature and life, a thread that leads to a sense of the broader and more encompassing reality, which in one form or another has always been a central concern of our religious traditions. It is this reality that, after all, contains and sustains us all in an unbroken web of interdependence.

Notes

1. See, for example, Edward O. Wilson, *The Diversity of Life* (Cambridge, Mass.: The Belknap Press, Harvard University Press, 1992).

2. "Appeal to the World," Global Forum of Spiritual and Paliamentary Leaders, Moscow, 1990. See Carl Sagan, "To Avert A Common Danger," *Parade Magazine*, March 1, 1992, p. 15.

3. Herman E. Daly and John B. Cobb Jr., *For the Common Good: Redirecting the Economy Toward Community, the Environment, and a Sustainable Future* (Boston: Beacon Press, 1989), 373–75.

4. See Arne Naess, "The Deep Ecological Movement: Some Philosophical Aspects," *Philosophical Inquiry* 8, no. 1–2 (1983):10–31.

Part I | A CALL TO AWAKEN

We have forgotten who we are.

We have forgotten who we are
We have alienated ourselves from the unfolding cosmos
We have become estranged from the movement of the earth
We have turned our backs on the cycles of life.

We have forgotten who we are.

We have sought only our own security
We have exploited simply for our own ends
We have distorted our knowledge
We have abused our power.

We have forgotten who we are.
Now the land is barren, the waters poisoned, the air polluted.

We have forgotten who we are.
Now the forests are dying, the creatures disappearing, humans
    are despairing.

We have forgotten who we are.

We ask forgiveness
We ask for the gift of remembering
We ask for the strength to change.

We have forgotten who we are.

—UN Environmental Sabbath Program

Before addressing our ecological situation from various religious perspectives, it might be helpful first to bring into focus the ethical and spiritual dimensions that underlie it and that seem necessary to ameliorate it. Timothy Weiskel, Paul Brockelman, and Steven Rockefeller provide just such a wake-up call from several different points of view.

In chapter 1, Professor Weiskel, who serves as director of Harvard Divinity School's Seminar on Environmental Values, introduces his essay with the story of Belshaz'zar's feast from the biblical Book of Daniel. He compares those attending the feast to modern "developed" economies that, by making gods of progress and unlimited growth and profit, have led to the present worldwide ecological crisis. Growthism, he tells us, is our most fundamental public religion and faith, a faith that has led to unprecedented and unsustainable resource depletion, biota destruction, and a general decline of (even) our own public health and welfare.

Weiskel believes that what is called for in such a situation is nothing less than a "radical theological revolution" that might lead, on the one hand, to a deeper questioning of the public faith in growth at any cost and, on the other, to a new awareness of human limits in the face of the possibility of ecological collapse. We need to alter our anthropocentric commitment to endless growth in favor of seeing ourselves as agents of a sustainable economy in balance with a larger nature of which we are a part. The handwriting is on the wall, although it is sometimes hard to read it, he tells us, "when your back is up against it."

Paul Brockelman is professor of philosophy and director of the Religious Studies Program at the University of New Hampshire. In chapter 2, Dr. Brockelman argues that if we are to alter our behavior toward nature, we must transform how we see it in much the same way that the conservationist and naturalist John Muir did. The ecological crisis we face is at root a spiritual issue, perhaps the spiritual issue of our time.

It is a spiritual issue, he thinks, because it has its roots in a human attitude toward nature and all of life, and it calls for a spiritual transformation in the way we conceive of it if ever we are to learn to treat it differently.

The particularly painful cultural, spiritual, and ethical situation in which we find ourselves is in fact "calling us to awaken from our benumbed and bewitched state." Religious awareness, then, may be absolutely indispensable in such an awakening, by integrating our lives into a wider reality beyond our own anthropocentric needs and by providing more appropriate visions of what life is about and how we ought to live it than we find in the consumer society.

Besides, Brockelman argues that people are hungering for a genuine religious reform and revitalization, ways of existing that are not only more meaningful in themselves but that might ground a more balanced way of being with nature than seems prevalent today. Such a spiritual awareness may help us to ameliorate our devastation of the earth by helping us to see that we are in fact born from it, live all our days within it, and ultimately will return to it. Seeing nature as epiphany means "seeing it with new eyes," as did John Muir.

Steven Rockefeller, professor and director of religious studies at Middlebury College, focuses his essay on reverence for life as the basis of an environmental ethic.

We live in a world, he says, in which almost one fifth of the world's population lives in poverty, in which conflict and violence are daily events, and in which the natural environment increasingly is being degraded and polluted if not in large measure destroyed. What are we to do about this? Dr. Rockefeller's central thesis is that reverence for life is the underlying principle that might help us achieve social justice, environmental health, and spiritual fulfillment.

A society's religious attitude toward life is fundamental in shaping how it develops and behaves toward both human and nonhuman alike. At the core of traditional religious attitudes, he tells us, lies a reverence for life that embodies feelings of awe and wonder in the face of the basic mystery of life. Following Albert Schweitzer, Rockefeller argues that there lies at the core of ourselves an inner realization that "I am life which wills to live in the midst of life which wills to live." All of life becomes sacred. Such compassionate reverence for life in all its forms can become the basis for human rights as well as the implicit rights of the natural world. If we can awaken to this profound sense of sacredness throughout nature, Rockefeller thinks, it can help us to recognize our responsibility to "do what we can to preserve and promote life."

I TIMOTHY C. WIESKEL

Some Notes from Belshaz'zar's Feast

King Belshaz'zar made a great feast for a thousand of his lords, and drank wine in front of the thousand . . . They drank wine, and praised the gods of gold and silver, bronze, iron, wood, and stone.

Immediately the fingers of a man's hand appeared and wrote on the plaster of the wall of the king's palace, opposite the lampstand; and the king saw the hand as it wrote. Then the king's color changed, and his thoughts alarmed him; his limbs gave way and his knees knocked together. The king cried aloud to bring in the enchanters . . . and the astrologers . . . Then all the king's wise men came in, but they could not read the writing or make known to the king the interpretation.

Then Daniel answered before the king ". . . you have praised the gods of silver and gold, of bronze, iron, wood, and stone, which do not see or hear or know, but the God in whose hand is your breath, and whose are all your ways, you have not honored."

"Then from his presence the hand was sent, and this writing was inscribed. And this is the writing that was inscribed: MENE, MENE, TEKEL, and PARSIN. This is the interpretation of the matter: MENE, God has numbered the days of your kingdom and brought it to an end; TEKEL, you have been weighed in the balances and found wanting; PERES, your kingdom is divided and given to the Medes and Persians." —DANIEL 5: 1,4,5 – 6,8,17,23 – 28

Throughout his writings Father Tom Berry, whose seminal theological reflections have given us new ways to think about God, has reminded us of two central themes: First, as human beings we are part of natural history, a larger evolutionary story, a geological story—indeed a cosmic story. Secondly, in that larger narrative we now find ourselves at a critical juncture—a key turning point in that entire narrative, the outcome of which will be determined in part by the beliefs we affirm through our daily behavior. Given his geological training and orientation Berry phrases this turning point in geological terms. We live, Berry says, in what can be described as the "terminal Cenozoic era." Before us we face the choice between the "technozoic" or the "ecozoic." In short, as humans we are an evolutionary outcome of natural processes, but our theology determines the character of our engagement with these processes, and it will thereby condition the outcome of the story itself.

One of the reasons some people find Berry's insights so disturbing is that it is uncomfortable to be reminded that we live simultaneously in multiple nested realities. When people point this out to us in everyday experience and we come to realize that they are right, we frequently feel we have been stupid, naive, duped, or misguided. The net result is that we feel sheepish about our previously bold assertions and a little humiliated by the whole experience.

The insights of geologists force this recognition upon our culture as a whole because geologists have a different sense of time than those of us pre-occupied with day-to-day events. Their professional perspective spans millions or billions of years. The evolution and extinction of entire species form but a small part of their purview. They are aware that the earth's history is nested within a larger narrative of cosmic evolution. Moreover, they are fully aware that within the earth's story are nested a whole series of more limited narratives involving the evolution of multicellular life, the development of life forms with central nervous systems, the development and demise of dinosaurs, the appearance of mammals, and eventually the evolution of human forms from the late Pleistocene onward. While much of this process has involved gradual, cumulative patterns of change, geologists are aware that there have been numerous abrupt discontinuities in the earth's history, marked by massive extinctions of numerous species.

In the face of the accelerating rate of ecological decline in our own experience, these large-scale scientific insights about the origins and cosmic context of human activity can prove to be disconcerting. Looking at the larger picture, for example, biologists reassure us that the invertebrates and microbial species are likely to survive our current epoch relatively un-scathed. Yet, if you are anything like me, this message provides small comfort when one begins to realize that the larger point is that *life as we know it* is undergoing massive extinction. More precisely, geologists, evolutionary biologists, and paleontologists are now reporting evidence in their professional journals that we are currently in the midst of a global "extinction event" which equals or exceeds in scale those catastrophic episodes in the geological record that marked the extinction of the dinosaurs and numerous other species.

At least two important differences exist between this extinction episode and those previously documented in the geological record. First, in previous events of similar magnitude the question of agency and the sequence of species extinctions have remained largely a mystery. In the current extinction event, however, we now know with a high degree of certainty what the effective agent of system-wide collapse is, and we have a fairly good notion of the specific dynamics and sequence of species extinctions. Second, previous events of this nature seem to have involved extraterrestrial phenomena, like episodic meteor collisions. Alternatively, the long-term flux of incoming solar radiation that results from the harmonic convergence the earth's asymmetrical path around the sun and the "wobble" on its axis also drive system-wide changes generating periodic advances and retreats of continental ice sheets in high latitudes. These too cause system-wide transformations and have precipitated extinction events in the past.

In contrast to these extraterrestrial or celestial phenomena that served as the forcing functions behind previous mass extinctions, the current extinction event results from an internally generated dynamic. The relatively stable exchanges between various biotic communities have shifted in a short period of time into an unstable phase of runaway, exponential growth for a small sub-set of the species mix—namely, human beings, their biological symbionts, and their associates.

The seemingly unrestrained growth of these populations has unleashed a pattern of accentuated parasitism and predation of these growing populations upon a selected number of proximate species that were deemed by them to be useful. This accentuated parasitism led to the creation of anthropogenic biological environments which, in turn, drove hundreds of other species directly into extinction—sometimes within periods of only a few centuries or decades. More significantly, however, this pattern of unrestrained growth and subsequent collapse has repeated itself again and again, engendering in each instance a syndrome of generalized habitat destruction and over time precipitating the cumulative extinction of thousands of species as one civilization after another has devastated its environment and dispersed its remnant populations far afield in search of new resources of plunder and squander.

For a variety of reasons—some of them apparently related to their religious beliefs—humans remain fundamentally ignorant of or collectively indifferent toward the fate of their fellow species, insisting instead that measurements of human welfare should be the only criteria for governing human behavior. Apparently, the "right to life" is effectively defined as the "right to HUMAN life." In system terms this anthropocentric belief in human exceptionalism has characterized past civilizations and remains no less dominant today. Scientists and techno-boomers alike promise us that technological miracles will save us from our rapidly deteriorating ecological circumstance and that no substantial sacrifice will be required of us. After all, "thanks to science" we have miracle crops, miracle drugs and miracle whip! What more can we hope for?

Well, the fact is we need a great deal more to survive as a society and as a species. In reality, we are just beginning to recognize the true immensity of the problem.

Consider, for example, the truly dramatic dimensions of our recent growth as a species. By recent here, I mean in evolutionary terms and in terms of the relatively long time scales required to engineer stable social adjustment to changing circumstance. In evolutionary terms, it took since the dawn of humanity to roughly 1945 for the human species to reach the total figure of about 2 billion people. That figure has more than doubled— indeed, nearly tripled—just since 1945. During the rest of our lifetime experts say that figure could well reach a total of 9 billion people if left to grow at projected rates.

Consider, as well, the overall ecological "footprint" of human expansion over the millennia, particularly as we have come to congregate in cities. Depending upon how one wishes to segment us from our biological relatives, humans have been around for roughly a million years or so. It is only in the last 1.2% of that history—roughly the last 12,000 years—that we have come to depend upon agriculture, and only the last 6,000 years or so that we have begun to transform our settlement patterns into urban concentrations. We are still in the midst of what might be called the "urban transition" in the human evolutionary experiment. It is not clear that the transition will be successfully achieved or that the human bio-evolutionary experiment will endure very much longer in evolutionary terms. Nevertheless, there is enough evidence available about the urban transition in human history to begin generating some general statements.

The new evidence of environmental archaeologists is especially sobering in this context. The history of cities has been associated with the history of repeated ecological disaster. The growth of cities has engendered rapid regional deforestation, the depletion of groundwater aquifers, accelerated soil erosion, plant genetic simplification, periodic epizootics among pest species and animal domesticates, large-scale human malnutrition, and the development and spread of epidemic disease. In many cases the individual elements of ecological decline have been linked in positive feedback processes, which reinforced one another and led to precipitous collapse of particular cities.

To overcome the limitations imposed by these patterns of localized ecological collapse, cities have historically sought to dominate rural regions in their immediate vicinity and extend links of trade and alliance to similarly constituted cities further afield. As arable land and strategic water supplies became more scarce and more highly valued, violent conflict between individual city-states emerged, leading in short succession to the development of leagues of allied cities and subsequently to the formation of kingdoms and empires with organized armies for conquest and permanent defense.

Even with the limitations of preindustrial technology, the results of these conflicts could be devastating indeed to local or regional ecosystems, particularly when victorious groups sought to destroy the ecological viability of defeated groups with such policies as scorched-earth punishment and the sowing of salt over the arable land in defeated territory. The ecological impact of warfare and the preparation for warfare has been devastating in all ages. C. S. Lewis's observation has proved sadly correct that "the so called struggle of man against nature is really a struggle of man against man with nature as an instrument."

Demographic historians have added further details to the picture of repeated ecological disaster painted by environmental archaeologists. Human populations have demonstrated again and again the long-term regional tendency to expand and collapse. These undulating patterns are referred to by demographers as the "millennial long waves" (MLW), and they appear to be manifest in both the old world and the new. Consider the regional data in figures 1 and 2.

Two patterns are discernible across all cases despite the considerable differences between each region. First, the human population is both highly unstable and highly resilient. That is to say, there is considerable variation in the amplitude of the population waves and therefore human populations cannot be considered stable in regional terms. Moreover, the population is resilient in the sense that it "bounces" back from demographic catastrophe with an even stronger surge in reproductive performance. The second phenomena of the MLW on the regional level is that the frequency between their occurrences is successively shortened. Thus, populations seem to be collapsing and rebounding at higher and higher levels more and more frequently as we approach the present.

When we move beyond the regional evidence to a global scale, another important pattern emerges. On this level of analysis it seems that human populations seem to expand in spurts, corresponding to the quantities of energy they are able to harness with their available technology. This may emerge as a new way of stating the Malthusian theory of population limit. Thomas Malthus focused on the relation of populations to their food supply and pointed out what while populations tend to grow exponentially

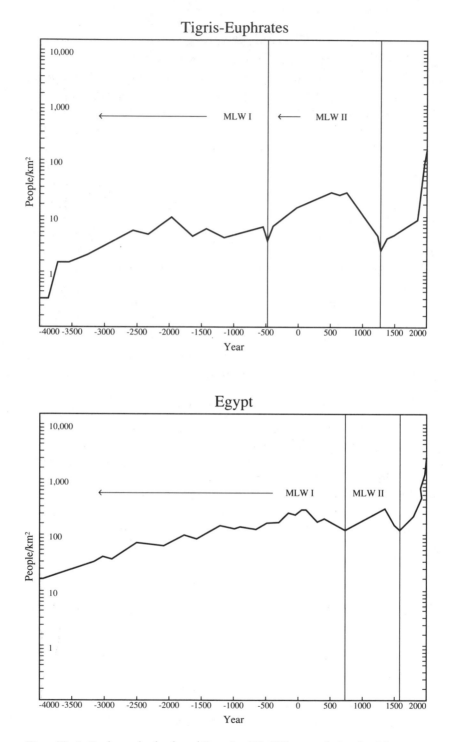

Fig. 1. Tigris-Euphrates lowlands and Egyptian Nile Valley population densities.

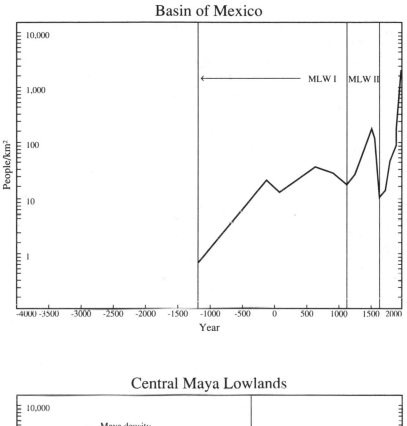

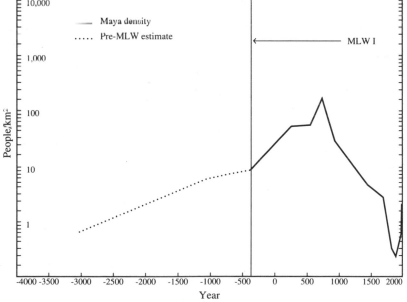

Fig. 2. Basin of Mexico and central Maya lowlands population densities.

the food supply tends to grow only arithmetically. As a result, populations are ultimately limited, according to Malthus, as their reproductive performance outstrips the food supply needed to keep them alive, and there are periodic widespread famines.

Since Malthus we have come to realize that "food" itself is really "energy"—a form of captured solar energy (i.e., kilocalories) that humans can assimilate to maintain themselves and do work. If we build upon this observation to reformulate Malthus's observation in terms of energy instead of food itself, we are probably close to a broad-level truth about the human species. Simply put, the Malthusian law can be restated in these terms: human populations tend to expand to the levels supported by the supplies of energy that they can mobilize with available technology.

The industrial era in world history marks an unprecedented period in human evolution history from this perspective. Never before have global populations experienced such high rates of growth for such sustained duration, reaching a worldwide climax with an average annual population increase of 2% during the decade from 1965 to 1975. The demographic historian Paul Demeny has described this extraordinary period quite succinctly:

It took countless millennia to reach a global 1700 population of somewhat under 700 million. The next 150 years, a tiny fraction of humankind's total history, roughly matched this performance. By 1950 global human numbers doubled again to surpass 2.5 billion. The average annual rate of population growth was 0.34% in the eighteenth century; it climbed to 0.54% in the nineteenth century, and to 0.84% in the first half of the twentieth. In absolute terms, the first five decades following 1700 added 90 million to global numbers. Between 1900 and 1950, not withstanding two world wars, an influenza pandemic, and a protracted global economic crisis, the net addition to population size amounted to nearly ten times that much.

As Dr. Demeny summarized the situation:

Clearly, viewed in an evolutionary perspective, the 250 years between 1700 and 1950 have witnessed extraordinary success of the human species in terms of expanding numbers, *a success that invokes the image of swarming* [emphasis added].

For demographic historians, then, it would seem that humans in the modern era are behaving much like a plague of locusts.

What is even more striking is that the pattern of distribution of this burgeoning population is one of rapid relocation into massive urban agglomerations. In 1700 less than 10% of the total world population of 700 million lived in cities. By 1950 a full 30% of the global population lived in cities. In North America the urban proportion of the population had reached 64% by that time, while in Europe it was 56%.

In 1700 only 5 cities in the world had populations of 500,000 people. By the turn of this century that number had risen to 43 cities in the world with populations of 500,000 or more. Of those, only 16 cities had populations over 1,000,000. By now, however—that is to say, in a span of under 100 years—there are nearly 400 cities that exceed 1,000,000. Moreover, the trend is accelerating, particularly in the Asian countries of the Pacific Rim. A recent report of the United Nations has indicated that "by the year 2000, the population of Dhaka is expected to double to 12.2 million; Bombay, Calcutta, Delhi, Jakarta, Karachi, Manila and Shanghai would each gain four million people; and Bangkok, Bangalore and Beijing [will each gain] three million." The numbers of mega-cities—those in excess of 10,000,000 people—will reach 21 by the turn of the century, with 13 of these in the Asia-Pacific region. By the year 2020 the report estimates that 1.5 billion more people will be living in Asian cities than live there today. This is the equivalent of creating a brand-new city of 140,000 people every day for the next thirty years.

The localized and global ecological costs of this seemingly unstoppable rush toward urban life are difficult even to imagine. While some techno-boomers and inveterate optimists suggest that newly planned cities might prove to be more energy and resource efficient, this kind of rapid urbanization has historically been accompanied by accelerated resource depletion, increased pollution and a decline of public health and welfare. In this large-scale process the "good life" for some has generally been purchased by the increased immiseration of many more and the nearly complete foreclosure on possibilities for a sustainable and stable livelihood of future generations.

Nowhere is this more apparent than in the massive transformation of the global food system in the last half century. The rapid growth of the world's population and its even more rapid urbanization since the end of World War II have meant that more and more food has had to be produced on a shrinking base of potentially arable land. While new land is still being brought into agricultural production, in the last decade or so the amount of arable land *per capita* has begun to decline on a global basis. This is a very ominous trend. Moreover, it seems to be an irreversible one.

So far the primary reason why this has not led in the short run to massive famine is that new, petro-intensive forms of agriculture have come to dominate global food production. Crops have been bred or engineered to respond to fertilizer inputs; crop losses and damage have been reduced by petrochemical pesticides and fungicides; competing weeds have been reduced by herbicides; and aridity problems have been overcome by using gas-driven pumps to extract fossil water from underground aquifers. In short, the increases in food production needed to support recent population growth and accelerated urbanization have been made possible through a more intensive use of non-renewable resources (topsoil, groundwater, and petroleum) in a farming system that generates ever more lethal side effects (pesticide residue poisoning, groundwater contamination by fertilizers, salinization of irrigated surfaces, agro-chemical "accidents" like Bhopal).

All of this has been accomplished on a rapidly declining crop genetic base, as indigenous varieties around the world are being displaced by varieties responsive to petro-chemical inputs. Never before in the history of humanity have so many people come to depend on so few plant species grown in such restricted regions and subsidized by the net destruction of such quantities of non-renewables. In just fifty years humanity has transformed global agriculture from a net source of captured solar energy into a net energy sink. We now face a situation as a species where our primary production system (agriculture) has become irretrievably dependent on a non-renewable (petroleum). At a time when atmospheric scientists tell us that erratic weather patterns and perhaps a changing climate regime may characterize the decades ahead, it seems likely that a stable global food supply will be harder to secure in the future than it has been in the past half century.

So much, then, for the environment and the transformations of it that we have wrought as a species. What about the "good" life? The obvious comment to be made in this context is that in our culture a desire to pursue "the good life" aggravates our momentous ecological crisis. Consumption patterns of the "Northern" countries and the "Western" countries are obscene by global standards, yet there is no apparent end in sight to the gluttony. Indeed, as citizens of the United States we have the right to "the pursuit of happiness" written into our constitution, and in culture the prevailing message is that happiness itself is inextricably linked to an ever greater consumption of material goods and energy.

In some cases, of course, individuals, households, and even entire communities have made great strides in efforts to reduce, reuse, and recycle. Nevertheless, the underlying economic logic of an economy based on unlimited growth remains largely unchallenged in public discourse. Not a single national political leader has been elected on a plank of steady-state economics. Indeed, I know of no candidate that would attempt to seek public office on a no-growth or a slow-growth platform.

Even if an exceptional candidate could be found to articulate this kind of appeal, such an effort would be laughed off the public stage. The reasons for this have as much to do with arguments about social justice as they do with shameless consumerism. After all, growth has become the only means that late capitalism has devised to cope with the increasingly evident problem of inequity. The promise of more tomorrow is at least partially successful in silencing those who object to the current distribution of goods and privileges. Political and social liberals are particularly easy to divert with this appeal. In general, they are well enough off in material terms not to question the fundamental system from which they benefit. When they go further and express concern for the welfare of those who have been left out of the system, they seem to be easily persuaded by the promise that more tomorrow will eventually do everyone good. After all, the system worked for them, why shouldn't it be thought to work for all others? In short, there is no significant debate between conservatives and liberals on the question of growth. Squabbles over relative rates or targeted sectors may occur, but they serve only to underscore the broadly shared consensus that growth itself is an unquestioned virtue and the only legitimate goal of all public policy.

This is why both national political leaders and Wall Street alike are driven with religious zeal to preach one or another variant of the gospel of growth. Ironically, those most victimized and marginalized by this organized system of accumulation are frequently persuaded by the proselytizers of growth that generalized expansion is their best available strategy for self-improvement. Thus it is that the poor become both the strongest justification of and supporters of the pro-growth evangelists.

Without any exaggeration, therefore, it is fair to say that in practical terms the most pervasive form of this religiously held belief in our day is that of *growthism* founded upon a doctrine of *techno-scientific salvation*. For the most part, the recent surge of "environmentalism" has not challenged this form of public religion. The fundamental belief is still that the earth and all it contains is constituted simultaneously as a treasure trove of raw materials and a repository for our wastes. When the absolute supply of resources is diminished, scientific discoveries and technical inventions, so it is believed, will save us from the constraints of absolute scarcity as new and more efficient production processes and waste treatment technologies are developed. Recycling itself is touted as a "growth industry" and a promising investment prospect on Wall Street. The sacred creed remains both pure and simple: *more is better; growth is good*. Anyone who expresses misgivings about this credo is soon taught through public rebuke and personal ridicule that it is blasphemy to question this golden rule of growthism.

We are confronted, therefore, in every respect with a growing problem. Given the pervasive character of the public faith in growth, it is impossible for the dominant forms of public religion to offer us a way out of our environmental crisis. From the vantage point of a systems ecologist or a "geologian," like Tom Berry, growth is the problem, not the solution. Yet the principle of continuous growth has achieved godlike status in the pantheon of modern religious icons.

When the high priests of public religion are asked, Can we survive?, their answer is emphatic: *Of course we can!* All we need is adequate investment incentives, a sense of determination, good ol' American inventiveness, and political will to make the "tough" decisions. One can hear the strains of "Onward Christian Soldiers" playing in the background as if we were "marching into war." The trouble with this is that the problems we are up against will no longer be solved simply with a new dose of messianic triumphalism.

In a narrow sense and in the short run we may succeed in "saving ourselves" from immediate manifestations of disaster, but it is essentially beside the point. The far more compelling question on a large scale and in the long run is, *will* we—as a species—survive? Not just theoretically, *can* we, but in a very practical sense, *will* we? This can only be answered by looking carefully at what we mean by "we" and what we mean by "survive." Growth evangelists and techno-scientific salvationists—like other fundamentalists—are regrettably silent and often sadly ignorant of the social dimensions of the changes required to answer this larger set of questions. Indeed, I would argue they are helpless in the face of such questions. Techno-boomers can do no more than offer us more of what got us into our sad circumstance in the first place.

It follows, therefore, that the only real chance we have of surviving as a species is through a radical theological revolution—that is, a thoroughgoing reexamination of those cultural beliefs we hold to most religiously. From the point of view of ecological sustainability, we have been weighed in the balance and found wanting. At current rates of growth and consumption our days have been numbered and the culture of growthism will be brought to an end whether we like it or not.

In some quarters this theological reformation is already underway. One can point to the most recent of a whole host of writers, from James Nash to Sally McFague or Jay McDaniel to Michael Fox, by way of supporting the point that church people and academic theologians are beginning to re-think concepts like "dominion," "stewardship," and "covenant" in terms that are more consistent with our contemporary ecological circumstance.

Further religious reflection on the relation between religious beliefs and the environment has also led to a reexamination of selected texts in the Judeo-Christian canon. Professor Theodore Hiebert at Harvard, for example, is currently in the process of retranslating the Yawist sources in the Hebrew Bible and will shortly publish an entirely new scholarly interpretation of the ecological setting of these early Hebrew scriptures.

Much of this effort is intended explicitly or implicitly to refute the assertion that the Judeo-Christian value system is somehow uniquely responsible for humankind's exploitative relationship with nature. Professor Lynn White leveled a stinging indictment at the dominant religious traditions of the West in just these terms in a 1968 article in *Science* magazine, and many of the writings from religious circles over the last twenty-five years have been largely defensive efforts protesting "no, it isn't so."

Other works from avowedly secular sources have served to let the Judeo-Christian tradition off the hook by pointing out that other ancient cultures were also devastating to their environments and seemed to similarly privilege human agency in the cosmic order of things. Thus, works like Donald Hughes's, *The Ecology of Ancient Civilizations* and a whole variety of subsequent ecological histories that it inspired have succeeded in spreading the blame fairly uniformly across all cultural traditions. Perhaps only the native American tradition has been spared a full-length ecological critique, but even here the burden of the evidence now being collected indicates that pre-Columbian civilizations did not represent the kind of ecological nirvana that some strains of contemporary environmentalism would have us believe.

These religious and cultural critiques are well intentioned and no doubt quite important in their own terms, but we need now to ask more fundamental questions. O.K., let's assume as given the two central points of all this recent scholarship: first, the Judeo-Christian tradition is more complex than one might think at first glance, allowing for, or indeed perhaps even encouraging, a far more ecologically sustainable approach to the environment than heretofore recognized. Secondly, virtually all other cultural traditions have in practice been equally exploitative of their resources. What of value, then, have we learned from all this? Have we learned to live more lightly on the earth? Have we effectively challenged the public theology of growthism in our day?

I think not. I would argue that what we need now is far more profound than proof-texting and retranslating our received traditions or launching yet other campaigns of cultural chauvinism in favor of one or another variant of the human achievement. What we need instead is a thorough-going reformation of our public theology of growthism.

We are all guests at Belshaz'zar's feast. On a global scale the handwriting is already on the wall for the culture of consumerism and its theology of growthism. Moreover, the meaning of this handwriting has been made plain. We are faced, as Tom Berry has suggested, with a choice between the "ecozoic" or the "technozoic." The question remains: will we behave like the king's "wise men"—the "enchanters" and the "astrologers"—and remain profoundly confused, or will we have the prophetic insight and the internal fortitude to challenge the public theology of our day?

The fundamental problem is that because of our patterns of growth our ecological impact as a species far outstrips our capacity to construct responsible communities of concern. We are just now beginning to monitor the radiological impact of the Chernobyl incident upon populations in nations far removed from the former Soviet Union. Less obviously but more insidiously, it is now possible to detect PCBs in the body fat of penguins in Antarctica. That is to say, the growing urban agglomerations around the world are already registering their ecological "footprint" in the snows of the last uninhibited continent. The mounting tragedy is that just as our collective behavior is registering a wider and wider ecological impact, our sense of effective community under stress is sharply shrinking.

A sense of moral compulsion cannot be imposed effectively from above, no matter how loudly it is preached from on high. Moral and ethical imperatives emerge spontaneously from a shared sense of community—a feeling that what "I" do or what "we" do matters to others within a community of which I wish to be a part. Our past record as a species is not encouraging in this regard. Historically, those considered to be *outside* the moral community have simply been ignored or—worse yet—legitimately persecuted in the name of the ethical principles of those *within* the boundaries of the recognized moral community. Clearly, our notions of what is *outside* and what is *inside* must change if we are to survive much longer as a human species in a wider biological community.

Environmental ethics, then, can be seen as an aspect of the more fundamental problem of community. In the time we have remaining can we fashion and believe in a collective sense of belonging to a global life process that transcends our home, our family, our class, our nation, and indeed our species? If our contemporary reactions to Somalia, Liberia, East Timor,

Haiti, Zaire, and numerous other "hot spots" around the world are any indication of what is to come, the signs are not entirely encouraging. Left to our default behavioral modes our effective sense of community seems to shrink in time of crises.

The discouraging fact is that throughout history religious identities and concepts of God have all too frequently been implicated in this pattern of inward-looking retreat from responsibility. In historical terms humans have not shown an ability to create and control stable ecological communities for very long, and many societies have accelerated their decline through an unreflective affirmation of outmoded religious beliefs. Unless exceptional leaders—religious and otherwise—can articulate a new vision of community and a compelling theory of human limit, we are likely to accelerate our demise by winning in the competitive struggle for dominance over all other species.

This, then, is what is meant by the need for a new theology. A theology is in essence a theory of human limit. Each culture and each age has had its own functional theology as the experience of human limit has varied through space and time. In our place and our time a forceful theory of human limit needs yet to be proclaimed with all the clarity of the prophetic pronouncements of old. The essential elements of such a theology are apparent: we live in a world we did not create and cannot control. This awareness inspires in whole people a feeling of humility, an enduring sense of wonder, and an abiding reverence for life itself. These sensibilities generate a profound sense of gratitude and motivate and orient our pursuit of truth, our struggles for justice, and our efforts to realize our potential as human beings. The outcome of our enterprise is *not* entirely in our hands, but the little that we do know about the world and our place within it allows us, nevertheless, to affirm meaning in the face of mystery.

This is where, in a modest way, I would say my own outlook departs most markedly from that of Tom Berry. The new narrative of cosmic "creation theology" that Tom Berry has inspired goes a long way to resituate the human species and its evolution in its proper natural history context; but there is a subtle danger in recounting this story, and it is simply this: we humans inevitably assign ourselves too large a role in the cosmic trajectory, as if our species were the goal or crowning achievement of evolution itself and perhaps of all cosmic process. In some formulations this perspective assigns to man a co-creative role with God for the unfolding of the future history of creation. This cannot be proved, but as with all fundamental beliefs it can be affirmed, declared, and proclaimed. In an effort to emphasize the important character of our responsibility as a species, it is tempting to emphasize the extraordinary power of the human species.

My own hunch is that such affirmations are a bit too grandiose. For my taste, the structure of this new creation narrative smacks too much of the old creation narrative wherein man was said to be made in the "image of God" and placed in a garden to tend and keep it and have dominion over it. In short, in some of its formulations the new narrative of creation theology can serve to engender and support an anthropocentrism which I feel is no longer credible and is potentially quite dangerous in sustaining the illusion that the future of the natural world is in *our* hands.

It is of course important to understand the beneficial ways in which we can interact with the environment, but it is equally important to understand the limits of human achievement in this regard and specifically what it is that we are not capable of doing. Announcing that we are co-creators with God in some process of cosmic self-realization is a bit like the rooster asserting that by crowing he makes the sun rise. If we are to be honest with ourselves and acknowledge what we have come to learn from science, we will need to start recognizing some real and palpable limits to the human prospect.

We are unlikely, for example, to be able to know enough to predict or perhaps even survive global climate change, so we had better build into our societies buffers and margins of collective safety that are much larger than any we have developed to date. We are unlikely to be able to win the co-evolutionary race with new and resurgent diseases, so we had better anticipate broad new public health strategies which are not predicated upon the "conquest" of disease.

We cannot regulate the earth's water cycle at will, particularly in the face of a potentially changing climate, so we should expect that limits on the availability and distribution of fresh water will pose limits on human expansion and industrial activity. Despite all our bio-technological wizardry in altering or modifying genetic material, humans have not "created" a single species. Instead we have only manipulated existing species for our perceived short-term benefit. Quite apart from the moral questions involved in the genetic manipulation of other species for human ends, it is unlikely that we will ever develop a predictive ecology that will be sophisticated enough to foresee the ultimate impact of introducing genetically altered species into the earth's complex ecosystem. We are not currently able to accomplish this kind of prediction for the thousands of new synthetic chemicals we introduce to the environment each year, and predicting the synergisms between these chemicals and life forms will probably prove to be beyond our reach.

Meanwhile, valuable genetic material in indigenous crop species and medicinal plants is being driven into extinction at rates that far exceed our capacity to catalogue the tragedy, let alone introduce new cultigens to take their place. We are unlikely to increase markedly the photosynthetic efficiency of the green leaf, so we had better begin to acknowledge that there are practical limits to the expansion of human numbers imposed by some photosynthetic process. Already it is calculated that roughly 40% of terrestrial photosynthesis is devoured by human beings, their animals or their industries. Even if we achieve the impossible and capture 100% of terrestrial photosynthate, the world's population cannot continue to double at its current rate without running into catastrophes of biblical proportions.

A sober assessment of our collective human limits suggests that even at our best we are perhaps not so co-creative as some new creation narratives would have us believe. This is not because we have merely been sloppy or asleep at the wheel. The problem goes deeper than this.

Human limits in the ecosystem stem from the basic fact that human societies and ecosystems operate most of the time on fundamentally different principles. As the noted ecologist Eugene Odum has phrased it, humans maximize for net production while ecosystems maximize for gross production. Ever since the advent of agriculture human societies have driven inexorably toward the logic of *more is better; growth is good*. Natural ecosystems operate on the contrasting principle: *enough is enough; balance is best*. The tension between these two principles is the ecological root of all evil. Humankind's repeated insistence upon trying to manipulate the larger ecosystem on the basis of its species-specific logic is the ecological equivalent of "original sin." The "sin" is original in the sense that it is built into our condition as humans. We can do no other. This aspect of the human condition cannot be overcome by pious good intentions to "do better" or earnest attempts to improve the efficiency of our maximizing strategies. It is these strategies themselves that are the source of the problem.

The only salvation from this condition is to step outside the strategy itself—to decenter ourselves and recenter our awareness around the logic of the larger system of which we are a part. This effort to recognize ourselves as part of a larger logic has been at the heart of both religious experience and scientific inquiry. Indeed, as William James pointed out nearly a century ago, the two endeavors are intimately linked:

. . . all the magnificent achievements of mathematical and physical science—our doctrines of evolution, of uniformity of law, and the rest—proceed from our indomitable
desire to cast the world into a more rational shape in our minds than the crude order of our
experience. . . . The principle of causality, for example—what is it but a postulate, an
empty name covering simply a demand that the sequence of events shall some day manifest
a deeper kind of belonging of one thing with another than the mere arbitrary juxtaposition
which now phenomenally appears? It is as much an altar to an unknown god as the one that
Saint Paul found at Athens. All our scientific and philosophic ideals are altars to unknown
gods.

If the story of Belshaz'zar's feast tells us anything, however, it is surely
that we should be wary of altars to unknown gods. It is as if we have been
warned that in our quest for what James calls "a deeper kind of belonging"
we should accept no substitutes.

Most disturbing of all, however, is the implication that even if we finally
get our theology right—as Belshaz'zar appears to have done—this fact is
not redemptive. There is no opportunity for confession, contrition, and
absolution—no assurance of forgiveness nor possibility of salvation. Consider the story's outcome. In spite of the fact Belshaz'zar has come to
understand the writing on the wall and appears to be genuinely chastised,
duly fearful, and appropriately grateful to Daniel, he is not spared. The
narrative records that after the feast he died that very night.

For those of us who have been steeped in Christo-centric theology, this
is disconcerting. We would prefer a more comforting closure. The message
is that even if we get around to reading the handwriting on the wall and
earnestly desire to change our ways, it is probably too late for those of us
who are already at Belshaz'zar's feast. If the story can be repeated often
enough and widely enough, perhaps others will benefit, but we will not be
spared.

It is perhaps precisely for this reason that notes on Belshaz'zar's feast may
well speak more directly to our culture's condition than the comfortable
stories of oneness with nature or the new narratives of belonging to a
cosmic creation. The message from researchers like Dennis Meadows and
others, including demographers, agronomists, atmospheric scientists, and
even some economists, is quite simply that the ecosystem will not support
or tolerate a global repetition of the development patterns characteristic of
the West and the North. In particular if the countries in Asia seek to
replicate historical patterns of Western resource use and energy exploitation and, in effect, praise as we have "the gods of gold and silver, bronze,
iron, wood, and stone," the fate of our species as a whole does not look
hopeful.

It is possible that Asian countries will learn from our sad environmental record just as we may have learned something from Belshaz'zar, but any realistic assessment of the probability of this occurring is rather small. The reason once again is simple: it is hard to read the handwriting on the wall when your back is up against it. Many Asian countries are experiencing the world's fastest rates of economic growth, and many more are promising ever greater growth as the only means of accommodating burgeoning populations. Already, as a recent international Academy of Sciences conference has demonstrated, Asian cities are becoming the largest concentrations of human immiseration on the planet. Public health officials are concerned about newly resistant forms of cholera emerging from Bangladesh, and it seems evident that many of Asia's forthcoming mega-cities will have to survive on radically reduced amounts of fresh water per capita by the turn of the century. Without massive and concerted efforts now to avert these circumstances it is difficult to see how epidemics and civil disorder on a large scale can be avoided.

Where is God in all of this? What attributes would such a concept possess? Are we prepared to believe in a God that seems poised to wreak such destruction, confusion and such massive suffering on the already poor and destitute? Whether or not academic theologians get their narratives reworked and their texts re-translated in time, I suspect that the effective theologies of the modern world are in for a radical and brutal transformation in the decades ahead. While from a comfortable distance we are on the verge of announcing that God's handiwork as manifest in the natural environment is a glorious harmonious whole, the mass of the world's humanity is about to endure a very different experience of human limit and divine presence. Metaphors of wanton destruction, vindictive revenge, and suffering innocence will probably be more consistent with their experience, and the concept of redemption may not take the form of worldly survival.

In short, while considering "God, the environment, and the good life" in the comfortable surrounding of New England, I suspect we should all be wary of overly domesticating the concept of God and re-creating him/her too much in our image. It is understandable that we all yearn for a new sense of belonging, a new sense of reconciliation with an alienated nature, but in our earnest and devout efforts to achieve this reconciliation we should be prepared to understand as well that for the mass of humanity other, more terrible concepts of God are likely to predominate for the foreseeable future. Unless we understand this and can learn to speak to this condition, we will have learned nothing from the notes we have taken at Belshaz'zar's feast.

With New Eyes

Seeing the Environment as a Spiritual Issue

In the last analysis, the psychological roots of the crisis humanity is facing on a global scale seem to lie in the loss of the spiritual perspective. Since a harmonious experience of life requires, among other things, fulfillment of transcendental needs, a culture that has denied spirituality and has lost access to the transpersonal dimension of existence is doomed to failure in all other avenues of its activities. —STANISLAV GROFF

Revisioning Life

Like many young men in their twenties, John Muir, who was later to become famous as a naturalist and conservationist, went through a period of profound turmoil and disorientation in which he struggled to find himself and his role in life. Pulled this way and that, he couldn't seem to discover who he was or was to become. Although he was "touched with melancholy and loneliness . . . and the pressure of time upon life," he was unable to settle upon a direction for his life and remained disoriented and mired in indecision.

It wasn't until an accident occurred to him in March 1867 that he was able to launch himself upon his career as a wilderness explorer. In a factory in which he manufactured agricultural implements of his own invention, a belt on one of the machines flew up and pierced his right eye on the edge of the cornea. He was blinded in that eye, and his left eye soon became blinded through nerve shock and sympathy. He was left in utter darkness. Unable to see, he tells us, "I would gladly have died. . . . My eyes closed forever on all God's beauty! . . . I am lost!"

After a careful examination, however, a specialist indicated that he would eventually see again, imperfectly in the right eye but normally in the left. What he needed to do was to remain for a month in a darkened room. He did that, all the while dreaming of wilderness such as Yosemite Valley in the Sierras. Finally, on an April day a little over a month after the accident, the remaining bandages were removed from his eyes and the shades from the windows. Beyond all hope and happiness, he was able to see the world again! He was, in fact, intoxicated by that resurrection of his sight. It was as if he were seeing everything anew, with new eyes as it were, fresh from the hand of God. The experience transformed him. With the awareness that he could find no happiness apart from wild nature and "that I might be true to myself," he reoriented his life to exploring that nature and advocating its conservation. "This affliction has driven me to the sweet fields," he said. "God has to nearly kill us sometimes to teach us lessons."[1] It was from this time that his continuous wanderings began. As he put it, "I bade adieu to all my mechanical inventions, determined to devote the rest of my life to the study of the inventions of God."[2]

The British philosopher John Wisdom tells an interesting story about religious knowledge or belief in a classic essay titled "Gods." It is a story that might help us better understand Muir's experience. Two friends, one a theist and one an atheist, return to a long-neglected garden of theirs. Weeds have sprouted up since they left, but in between the weeds they find a few of the old plants still surprisingly vigorous. Having inspected the entire garden, the theist comes to the conclusion that an invisible gardener has been taking care of it, whereas his atheist friend concludes that there has been no invisible gardener. Both agree about all the facts: gardens need sunlight, water, fertile soil, and so on. In fact we can even imagine that the friends carry out a thorough study to ascertain all the facts that might influence and determine any possible garden, and they reach total agreement about them. Thus, Wisdom seems to be saying, their varying beliefs concerning the existence of an invisible gardener who tends the garden is simply not a factual or empirical hypothesis that can be demonstrated experimentally. It would seem, then, that both the theist's belief in an invisible gardener and his atheist friend's contradictory belief that there is no such gardener are more like ways of "seeing" the garden as a meaningful whole than like empirical hypotheses that are confirmed or disconfirmed by any possible facts concerning gardens.[3]

In this sense, spiritual understanding is more like suddenly seeing the famous gestalt figure meaningfully either as a vase or as two faces than like constructing an empirical hypothesis or a deductive syllogism. Religious faith and insight provide an overarching interpretive understanding of life as a *meaningful* whole, including our own role and destiny within it.[4] John Muir's experience, then, was a religious revisioning, a revisioning that transformed not only how he saw nature, but also how he envisaged his role within it as a naturalist and conservationist. As we have seen, he changed how he lived because of it.

Pushed by the stultifying and painful spiritual condition in which he had been living and transformed by the shock of his temporary blindness, Muir came to see nature with the amazed eyes of a child again and to understand his own role within it in a new way. In his early essay "Nature," Ralph Waldo Emerson had described such a transforming vision this way: "Few adults can see nature. Most persons do not see the sun. At least they have a very superficial seeing. The sun illuminates only the eye of the man, but shines into the eye and the heart of the child."[5] Muir's wonder at the extraordinary miracle of life, at the incredible epiphany it manifested, touched him to his core and enabled him to find his authentic orientation in life. In traditional religious terminology, he became spiritually reoriented because he discovered his own connection to a broader, sacred reality and community to which he belonged, a reality that permitted him to see how he might live more deeply and meaningfully than hitherto. He put it this way in his journals:

The man of science, the naturalist, too often loses sight of the essential oneness of all living beings in seeking to classify them in kingdoms, orders, families, genera, species, etc., taking note of the kind and arrangement of limbs, teeth, toes, scales, hair, feathers, etc., measured and set forth in meters, centimeters, and millimeters, while the eye of the Poet, the Seer, never closes on the kinship of all God's creatures, and his heart ever beats in sympathy with great and small alike as "earth-born companions and fellow mortals" equally dependent on Heaven's eternal love.[6]

His spiritual transformation, then, wasn't so much a shift in how he thought about things as a shift in how he looked at them, how he felt about them, and how he actually acted and behaved toward them. He found his way in life by finding his way home to nature.

All of us at various times have touched the spiritual and moral condition at a deep level of seriousness. Perhaps it happened during a divorce, the death of a parent, hitting bottom after a serious addiction, the loss of a job on which one depended financially or emotionally, the outbreak of war, or some other trauma that led to a disintegration of one's familiar and everyday way of seeing things.

Such spiritual reorientations as that of Muir, of course, are not limited to individuals alone. Historians and scholars of various kinds have long been aware that human cultures also occasionally undergo such transformations in how they envisage life as a meaningful whole and how they picture the purpose and role of humans within it. To find examples of such paradigm shifts in the fundamental worldview of our own culture we would have to go back to the cultural revolution constituted by the replacement of fertility goddesses with male warrior gods after 2500 B.C.E; the shift from polytheism to "radical monotheism" (to use H. Richard Niehbur's trenchant phrase) in early Jewish history; the change to Christianity in fourth- and fifth-century Rome; and the startling transformation of the by-then traditional European Christian culture into what we now call "modernity" or "the modern world" in the seventeenth, eighteenth, and nineteenth centuries.

In the face of the ecological difficulties avalanching down upon us, it may be that all of us, like Muir, will be forced to reevaluate how we "see" nature and change our behavior toward it. Many observers of our contemporary world, in fact, argue just that and that such a reflective reevaluation and reorientation of our lives will entail a digging down to the foundations of our ultimate faith in life. In other words, getting our ecological bearings may first entail getting our spiritual bearings in life by finding our way back to our home in nature.

A Bird's-eye View of Our Ecological Situation

It is probably safe to say that the present environmental state of the world constitutes the most serious threat to the biosphere since the origin of life on earth. It is also safe to say that the environmental crisis is not only a threat but also a situation that will not be easily overcome and that will haunt us for the foreseeable future. In the recent words of Pope John Paul II, "our problems are the world's problems and burdens for generations to come."

Indeed, the all too familiar phrase "ecological crisis" may be too feeble a way to put it. It is becoming increasingly clear to a number of observers that this is a crisis of the whole life system of the modern industrial world, one that affects both nature and the human culture it supports and sustains. Indeed, we seem to be living in a time in which we are witnessing not only breakdowns in the natural systems of the biosphere into which we have intruded with our economic and technological "progress" but also breakdowns in important parts of those economic, political, and cultural systems themselves. It seems increasingly clear that the familiar model of reality that hierarchically separates the human from the rest of life, or human cultures from nature, is both false and destructive of that wider nature.

In fact, contemporary science clearly shows that everything that has emerged on earth has emerged from and within nature as a whole. From this point of view, the economic, social, moral, and spiritual decay that is often manifested in our present world is not something that lies "outside" nature but is a biocultural development "within" it. Such cultural decay, then, is just one more manifestation of ecological disturbances and difficulties introduced by the modern industrial world. With all its obvious benefits, that modern industrial society that has so devastated our natural environment seems increasingly to be devastating us as well.[7] Putting the same thing another way, it would seem that to the degree that we have lost our sense of being rooted in a deeper and more encompassing natural order or reality, we have become spiritually, morally, and ecologically disoriented.

It would seem, then, that the avalanche of environmental issues we are currently witnessing around the globe calls for long-term consideration of how we are living and how that affects both the environment and ourselves rather than merely short-term technological "fixes." And yet such long-term consideration is difficult for all of us precisely because we are so caught up in the pursuit of short-term economic and political "success." As Harvard theologian Gordon Kaufman has put it in his most recent book, "The organization of human economic life into institutions geared to satisfying human needs and wants . . . , and of political life into nation-states, prevents us from directing our concerns and energies toward the larger world beyond our human-centered interests, and working for the common good of all creatures."[8]

Yes, But Is the Environment a Spiritual Issue?

Although certainly in part economic, demographic, and political in nature, the earth's ecological deterioration is at heart a matter of human attitudes toward the earth and life in general, attitudes that of course affect how we behave toward it. Thus, it would seem to constitute a spiritual crisis involving our moral and spiritual attitudes toward nature and, in fact, life as a whole. It may call for spiritual reflection on what we consider to be of ultimate importance in our lives and how we think we ought to live in the light of that and moral reflection on how we understand and relate to nature.

But why? Is nature and our behavior toward it in any way a spiritual question? And why is it that the environment, which previously had rarely been thought to be such a spiritual issue, has in fact suddenly become so for so many today? I think there are basically four reasons for this remarkable shift.

First of all, there is increasing recognition that a spiritual attitude toward nature has contributed to the increasingly dangerous environmental destruction and collapse we now see all around us. Newton, of course, thought of nature as an intricate machine fashioned by a designer God but running on its own according to the laws of mechanics. Since then, due to the ensuing industrial revolution, we seem to have totally commodified nature. We conceive of it as mere stuff stripped of any intrinsic value before it is forcibly extracted from the "wild" (meaning uncontrolled) and brought into the human economy—a "natural resource," as we put it, ready to be transformed industrially into useful products to improve the human condition. Aldo Leopold put it this way in his *Sand County Almanac*: "We abuse land because we regard it as a commodity belonging to us. When we see land as a community to which we belong, we may begin to use it with love and respect."[9]

Far from being a scientific or neutral hypothesis, this view of nature as a commodity put here simply for our use is itself an interpretive understanding, a way of seeing nature in much the same sense that John Wisdom's atheist (as well as the theist) brings a point of view to the garden beyond factual hypothesis. It is, then, a perspective on nature and life itself, a spiritual vision (if one can use that term for a point of view that denies the possibility of "spiritual" perspective at all), which of course is (or ought to be) quite different than, for example, a Jewish or Christian or Buddhist perspective. Many believe that a root cause of our violent destruction and transformation of nature lies in just such a modern perspective, which strips nature of any intrinsic value, not to mention epiphany. It hardly seems possible that without that modern way of seeing it we could have violated nature in quite the way we have and to the extent that we have. In this sense, then, the environment would seem to be unavoidably a spiritual issue.

But second, connected to (or perhaps embracing) this materialistic view of nature is a materialistic conception of "the good life." This is a thin vision of life as a whole that has flowered in the twentieth century, a vision suggesting that the central thrust and significance of our lives consists of the accumulation of capital or material goods. As the bumper sticker puts it, "Whoever has the most things when he dies, wins!" It is almost as if the accumulation of goods and the kind of intensive attention their production and consumption entails shields us from death, as indeed it does seem to do to a certain degree.

This is a closed and mean conception of life, which not only thinks of nature as put here simply for our enjoyment but anthropocentrically places human life at the center of everything (the entire universe!). Having accomplished that marvelous trick, it then makes morality radically relative to the wants and desires of a particular group or, of course, even each individual. Its conception of progress, of course, is the continuing expansion of economic demand and the industrial production to achieve it. This is what some of the authors in this book call "growthism." If, as Paul Tillich used to insist, religion means simply a group's "ultimate concern," then growthism would seem to be our religion and the gross national product our god. But all of that exacerbates the destructive and violent intrusion of human culture into nature.

It also leads to what Vaclav Havel has called "a demoralized culture" in which ethical ideals are simply reduced to the dreams of the consumer society or the lonely individuals who inhabit it. Finally, it brings about a culture of spiritual collapse in which there is no vision of a wider or deeper reality of which we are part than our own desires and dreams. That, of course, is precisely a spiritual emptiness or nihilism. Indeed, some commentators believe not only that this sense of spiritual emptiness is growing but that it is leading to more profound social fragmentation and violence, including, quite evidently, the destruction of the family. As Richard Eckersley put it recently, our materialistically oriented consumer society is increasingly failing

. . . to provide a sense of meaning, belonging, and purpose in our lives, as well as a framework of values. People need to have something to believe in and live for, to feel they are part of a community and a valued member of society, and to have a sense of spiritual fulfillment — that is, a sense of relatedness and connectedness to the world and the universe in which they exist.[10]

David Bollier has argued in a recent issue of *Tikkun* that we must come to grips with this ethical and spiritual emptiness: "The truth is, Americans in the late twentieth century need more than the First Amendment and its case law to bind them together. They need a new cultural covenant with each other that can begin frankly to address the spiritual void in modern secular society."[11]

At any rate, it is this materialist vision of prosperity, progress, and the good life that seems so rampant in our culture and so destructive to the environment. It is surely unworthy of free men and women. But that religiously oriented and practicing free men and women have shown little interest in questioning such a collapsed spiritual point of view from the perspective of their faith traditions—at least until recently—seems absolutely astounding! Can we seriously believe that God favors such materialism and growthism, especially insofar as they have brought about an unprecedented assault on creation itself? This is not an objection to free-market economies. But to connect such market economies to the (albeit myopic) spiritual vision that the end and purpose of life (the good life) is a surfeit of material accumulation and security hardly seems worthy of such faiths. Are our religious traditions really so threadbare and lacking in imagination, so timid, that they reduce their visions of the life of faith to that of the consumer society?

On the one hand, some deconstructionists have argued that the lack (and from their point of view the impossibility) of any deeper or more encompassing vision of life is precisely a problem that cannot be overcome. It may be, on the other hand, that the very material culture that has led to such painful nihilism and that has brought such horrendous devastation upon the environment will, for those very reasons, inevitably lead us beyond its myopic perspective. "Despite claims by social critics like Lyotard and Frederick Jameson that our society reflects the absence of any great integrating vision or collective project, the great collective project has, in fact, presented itself. It is that of saving the earth—at this point, nothing else really matters."[12]

The need for a serious ethical response to nature and the environmental situation in which we find ourselves is, I believe, a third reason that the environment is a spiritual issue. If we are to change our abysmal behavior toward the environment, we will need more than scientific analysis and social legislation: we need a moral perspective and code that can help to change that behavior. As Senator Gaylord Nelson put it at a recent interfaith conference in Virginia, "The harsh reality is that no war, no revolution, no peril in all of history measures up in importance to the threat of continued environmental deterioration. . . . The absence of a pervasive, guiding conservation ethic in our culture is the issue and the problem. It is a crippling if not, indeed, a fatal weakness."[13]

I find it interesting that until recently, "ethics" was a field in philosophy limited to human interaction. This was a reflection of our anthropocentric view that human beings lie outside or beyond nature. It is only humans who feel ethical obligation, and such obligation is directed only toward other human beings. In short, this view implied that we have no ethical obligations to individual plants and animals, never mind bioregions or nature as a whole because these entities have no "feelings" about how we treat them. I suppose this would be like a collection of trees agreeing—if they could—that unless you have roots, bark, and leaves you're beneath any sympathy or consideration. Fortunately, this myopic limitation of moral responsibility to human beings is now being seriously questioned.

Our system of dealing with nature simply as a collection of commodities put here for our privileged use seems to have failed or at least is in the process of failing. Aldo Leopold suggested (and others have followed his lead) that the only way to overcome our destructive treatment of nature is to treat it ethically, that is, as a *community* to which we belong. "All ethics so far evolved," he writes, "rest upon a single premise: that the individual is a member of a community of interdependent parts."[14] In other words, the new ecological ethic must extend our moral obligations to the larger community of nature to which we belong and that ultimately constitutes a single, interdependent web of entities, just as John Muir argued.

But how are we are to jump from the biological "is" to the ethical "ought," from theory to actual behavior, from information to wisdom, from understanding to passionate caring? Science, especially the science of ecology, tells us that nature is a single community; how, then, do we *actually* come to love and respect it? The answer, as many now argue, lies in going beyond its *utility for us* to a feeling-awareness of its *intrinsic value* in and of itself. As Arne Naess has put it, "The well-being and flourishing of human and nonhuman Life on Earth have value in themselves (synonyms: intrinsic value, inherent value). These values are independent of the usefulness of the nonhuman world for human purposes."[15]

But in order to do that, must we not, in fact, feel a reverence for this larger community to which we belong, must we not come to see it differently, in much the same way that Muir came to see it after the restoration of his sight, as God's holy creation? In other words, does not an effective ecological ethics, if it is to be more than an abstract set of principles, rest on a spiritual attitude toward the larger natural community to which we belong? Mustn't a serious and effective ecological ethics be grounded and founded upon a deeper and wider spiritual vision of life than seems available in the modern consumer societies (which interestingly enough have developed the very notion of utilitarian ethics)?

In a series of letters he wrote to his wife, Olga, in 1982 while imprisoned by the Czechslovakian Communist government, Vaclav Havel (later president of the newly founded Czech Republic) indicated that people living within modern industrial societies, whether capitalist or communist, all too often envisaged no wider, more encompassing, or more significant reality beyond their own needs and desires. Such a worldview, he thought, constitutes a kind of "demoralization."

We live in an age in which there is a general turning away from Being: our civilization, founded on a grand upsurge of science and technology, those great intellectual guides on how to conquer the world at the cost of losing touch with Being, transforms man its proud creator into a slave of his consumer needs. . . . A person who has been seduced by the consumer value system, whose identity is dissolved in an amalgam of the accoutrements of mass civilization, and who has no roots in the wider order of Being, no sense of responsibility for any higher reality than his or her own personal survival, is a demoralized person and, by extension, a demoralized society.[16]

The result of this inability to envision ourselves as a part of a larger reality, whether divine or merely natural, has led, then, to a demoralized culture in which all too often we see ourselves as disembodied intellects who are "outside" or "above" nature and thus free to manipulate it for our own selfish ends. In short, it has led to a collapsed spiritual vision and moral stance in which, as we saw, nature is "beneath" us and not even thought to have "rights" or to call for moral obligations on our part. It is seen to be a mere "stuff" put here for our enjoyment, simply the backdrop for the drama of the progressive unfolding of human history. Vice President Gore, in his recent book, *Earth in the Balance*, puts it this way: "Believing ourselves to be separate from the earth means having no idea how to fit into the natural cycle of life and no understanding of the natural processes of change that affect us and that we in turn are affecting. It means that we attempt to chart the course of civilization by reference to ourselves alone. No wonder we are lost and confused."[17]

This lack of awareness and appreciation for any "wider order of Being," as Havel put it, this "demoralization," has its roots, as we saw, in the spiritual worldview (how we "see" nature and life as a whole) that lies at the heart of our modern industrial cultures. Spiritual fire must be fought with spiritual fire. Any ethics on which we might pin our hopes of changing human behavior toward the environment must rest, ultimately, on a spiritual vision that transforms us, as it did Muir, and permits us to experience it in a reverential way as intrinsically valuable. If we are to change our behavior toward nature, if we are to act ethically toward it, we must look at it and our place within it differently. As Havel put it on another occasion, "The challenge offered by the post-Communist world is merely the current form of a broader and more profound challenge to discover a new type of self-understanding for man. . . . we must discover a new relationship to our neighbors, and to the universe and its metaphysical order, which is the source of the moral order."[18]

In a recent Fourth of July speech in Philadelphia, Havel developed this theme further by grounding respect for others, including nature, in a more profound spiritual vision.

Politicians at international forums may reiterate a thousand times that the basis of the new world order must be universal respect for human rights, but it will mean nothing as long as this imperative does not derive from the respect of the miracle of being, the miracle of the universe, the miracle of nature, the miracle of our own existence. Only someone who submits to the authority of the universal order . . . can genuinely value himself and his neighbors, and thus honor their rights as well.[19]

So this is another reason that the environment is a spiritual issue: any ethical approach ultimately rests on a spiritual way of seeing it. It was Albert Schweitzer, of course, who based his ethics on his spiritual sense of reverence for all life. He gained that reverence, he tells us in his autobiography, through an actual spiritual experience he had while crossing a river through a herd of hippopotamuses in Africa: "I am life which wills to live, in the midst of life which wills to live."[20] That experience of a reverence for life led Schweitzer to his explicit ethics, an ethics that parallels that of Aldo Leopold insofar as it links ethics and the wider natural community to which we belong.

The great fault of all ethics hitherto has been that they believed themselves to have to deal only with the relations of man to man. In reality, however, the question is what is his attitude to the world and all life that comes within his reach. A man is ethical only when he devotes himself helpfully to all life that is in need of help. Only the universal ethics of the feeling of responsibility in an ever-widening sphere for all that lives—only that ethic can be founded in thought. The ethic of the relation of man to man is not something apart by itself: it is only a particular relation which results from the universal one.[21]

Forth and finally, there is a hunger across the land for a genuine spiritual vision and life beyond the constricted and narrowing confines of the consumer society. And where might we find such an encompassing sense of life as a meaningful whole if not within the universe or creation as a whole? That is, we are inextricably tied to both the earth community and the larger universe from which it has evolved. Can what Havel calls a wider vision of reality be other than Being or Reality itself, that is, the whole fecund fifteen-billion-year unfolding of the universe? There is a widespread thirst for "reality," especially on the part of our young. What could possibly be more real than reality itself, whether it be called nature, God, life, or the originating mystery that shines through that nature?

The fourth reason for thinking that the environment is a spiritual issue, then, lies in the fact that environmental concerns may make possible a genuine religious reform and renewal, not in the sense of dogma but in the sense of experiencing with John Muir the epiphany that nature exhibits. It would seem that our time is calling us to awaken from our benumbed and bewitched state to a wonder at and reverence for the astonishing, miraculous, and mysterious creation of which we are a part. The whole world seems to arise in a mysterious emptiness. Reality is a transcendent but astonishing and holy power-to-be, an ever-flowing river of grace, a jaw-dropping gift of infinite giftedness. The gulf between nothing and something is filled with wonder, gratitude, and love of everything!

In a report to its 1991 General Assembly in Canberra, the World Council of Churches expressed just this sense of the sacredness of nature in its own Christian imagery.

Instead of a king relating to his realm, we picture God as the creator who "bodies forth" all that is, who creates not as a potter or an artist does, but more as a mother. That is to say, the universe, including our earth and all its creatures and plants, "lives and moves and has its being" in God (cf. Acts 17:28), though God is beyond and more than the universe. Organic images seem most appropriate for expressing both the immanence of God in and to the entire creation as well as God's transcendence of it. In the light of the incarnation the whole universe appears to us as God's "body."[22]

The Unfurnished Eye

Yes, the environment is a spiritual issue. For that reason, religious consciousness and perspective may be indispensable in ameliorating our present situation by helping us to integrate ourselves in a wider (and surely wiser) natural reality and by suggesting alternative conceptions of "progress" and the "good life." As the 1992 State of the World report of the Worldwatch Institute puts it, "With current notions of economic growth at the root of so much of the earth's ecological deterioration, [what is called for is] a rethinking of our basic values and visions of progress."[23]

Unless and until we change our basic attitudes toward nature (and the relationship of God to nature) and our conceptions of what constitutes progress and the good life, it may be that further environmental devastation will be inevitable. What is called for, then, is a vision of how to live appropriately in the face of the truth of nature. We don't need to save the world; we need to love it. As Father Zosima puts it in Dostoevsky's The Brothers Karamazov, "Love all of God's creation, the whole and every grain of sand in it. Love every leaf, every ray of light. Love the animals, love the plants, love everything. If you love everything, you will perceive the divine mystery in things."[24]

And whether the extraordinary unfolding of life in its myriad forms is called God's creation, the Tao, the body of the Buddha, or just plain nature is not as important as perceiving it once again with a child's wide-eyed amazement. Rachel Carson certainly knew that.

A child's world is fresh and new and beautiful, full of wonder and excitement. It is our misfortune that for most of us that clear-eyed vision, that true instinct for what is beautiful and awe-inspiring, is dimmed and even lost before we reach adulthood. . . . I should ask that . . . each child in the world [develop] a sense of wonder so indestructible that it would last throughout life, as an unfailing antidote against the boredom and disenchantment of later years."[25]

In Emily Dickinson's marvelous phrase, to perceive it with "an unfurnished eye" is to see it as the epiphany it truly is; it is to see and feel the sanctity of life in all its wondrous forms. As was the case with John Muir, that just may be the way for us to find our ecological way home.

Notes

1. Linnie Marsh Wolfe, *Son of the Wilderness: The Life of John Muir* (New York: Alfred Knopf, 1946), 104, 105.

2. William Frederic Bade, *The Life and Letters of John Muir* (Boston: Houghton Mifflin Co., 1924), 155.

3. John Wisdom, "Gods," in *Religion from Tolstoy to Camus,* ed. Walter Kaufmann (New York: Harper Torchbooks, 1964), 391–406.

4. For more on this view of religious understanding, see Paul Brockelman, *The Inside Story: A Narrative Approach to Religious Understanding and Truth* (Albany, NY: State University of New York Press, 1992).

5. Ralph Waldo Emerson, "Nature," in *The Portable Emerson*, ed. Carl Bode (New York: Penguin Books USA, 1981), 10.

6. *John of the Mountains: The Unpublished Journals of John Muir*, ed. Linnie Marsh Wolfe (Boston: Houghton Mifflin Co., 1938), 434.

7. In a very interesting article, Robert Kaplan argues that ecological disruptions will constitute the fundamental issue for our foreign relations in the years ahead. See Robert D. Kaplan, "The Coming Anarchy," *Atlantic Monthly*, February 1994, 44 ff.

8. Gordon Kaufman, *In Face of Mystery: A Constructive Theology* (Cambridge, Mass.: Harvard University Press, 1993), 313.

9. Aldo Leopold, *A Sand County Almanac* (New York: Oxford University Press, 1949), viii.

10. Richard Eckersley, "The West's Deepening Cultural Crisis," *Futurist*, November/December 1993, 10.

11. David Bollier, "Who 'Owns' the Life of the Spirit?" *Tikkun*, January/February, 1994, 89.

12. Suzi Gablik, *The Reenchantment of Art* (New York: Thames and Hudson, 1991), 26.

13. Senator Gaylord Nelson, keynote speech at the 1993 Interfaith Launching of Earth Week conference, quoted in *EcoLetter*, North American Coalition on Religion and Ecology (NACRE), Washington, DC, 1993, p. 4.

14. Leopold, *A Sand County Almanac*, 203.

15. See Arne Naess, "The Deep Ecological Movement: Some Philosophical Aspects," *Philosophical Inquiry* 8, no. 1–2 (1983): 10–31.

16. Vaclav Havel, *Letters to Olga*, trans. Paul Wilson (New York: Knopf, 1988), 365–66.

17. Al Gore, *Earth in the Balance* (New York: Houghton Mifflin Co., 1992), 162–63.

18. Vaclav Havel, "The Post-Communist Nightmare," *New York Review of Books*, May 27, 1993, 10.

19. Vaclav Havel, "The New Measure of Man," *New York Times*, op-ed page, July 8, 1994.

20. Albert Schweitzer, *Out of My Life and Thought*, trans. C. T. Campion (New York: Henry Holt & Co., 1933), 186.

21. Ibid., 188.

22. "Liberating Life: A Report to the World Council of Churches," in *Liberating Life: Contemporary Approaches to Ecological Theology*, ed. Charles Birch, William Eakin, and Jay B. McDaniel (Maryknoll, N.Y.: Orbis Books, 1990), 279.

23. Sandra Postel, "Denial in the Decisive Decade," in *1992 State of the World*, ed. Lester R. Brown (New York: W. W. Norton & Co., 1992), 4.

24. Fyodor M. Dostoevsky, *The Brothers Karamazov*, trans. R. Pevear and L. Volokhonskky (San Francisco: North Point Press, 1990), 319.

25. Rachel Carson, *The Sense of Wonder* (New York: Harper & Row, 1965), 42–43.

3 STEVEN C. ROCKEFELLER

The Wisdom of Reverence for Life

One of the most popular images to emerge from nineteenth-century American painting is "The Peaceable Kingdom" as portrayed in the work of Edward Hicks. The artist did over fifty variations on this theme, all inspired by his Quaker faith and the celebrated eleventh chapter of the Book of Isaiah. The prophet Isaiah's vision describes the reconciliation, peace, and happiness that will pervade all of nature as well as human society when the Messiah appears and the earth is governed by God's wisdom and righteousness. From Hicks's Quaker perspective, his paintings are also an image of the spirit of Christ and the transforming power of the Inner Light.[1] This vision is especially significant for our time because it is the expression of an attitude of profound reverence for all life, and it suggests that the fate and fulfillment of humanity and nature are intimately connected.

Consider the opening lines of Isaiah's description of paradise regained: "The wolf shall dwell with the lamb, and the leopard shall lie down with the kid, and the calf and the lion and the fatling together, and a little child shall lead them" (Isa. 11:6). The ethical theme of the prophet's vision is clearly stated in verse 9: "They shall not hurt or destroy in all my holy mountain"; Isaiah further explains that in the Messianic age when the Kingdom of God is established on earth no creature will do violence to another: "for the earth shall be full of the knowledge of the Lord, as the waters cover the sea" (Isa. 11:9). Here the knowledge of God is equated with a wisdom that involves an appreciation of the intrinsic value of all life and a compassion that seeks to prevent suffering and to create shalom, a joyful community of all living beings.[2]

Two thousand six hundred years after Isaiah we find ourselves in a world in which an extraordinary industrial and technological revolution has brought affluence to some but has not created anything like the universal well-being that many imagined it would. Over one billion people live in absolute poverty. Conflict and violence are commonplace throughout most of the world, and our own society has a deadly obsession with guns. The natural environment, upon which all life is dependent, is being dangerously degraded, and the biodiversity of the planet is being destroyed. All of these problems are intensified by an unchecked global population explosion. A sense of meaninglessness and hopelessness is widespread among vast numbers of young people, many of whom are trapped in urban wastelands. These interrelated problems form the dark side of the twentieth century's frantic quest for material progress and worldly happiness. They threaten to overwhelm human civilization unless some radical changes are made.

Perhaps Isaiah and Edward Hicks glimpsed a truth that can help us, both as individuals and as communities. Their vision of shalom and reverence for all life involves the idea of a better way, from which we can learn even though we do not live in paradise and lions should not be expected to lie down with sheep. The argument of this essay is that reverence for life is a supreme ethical principle essential to the realization of justice, peace, and environmental protection on earth, and it constitutes the spiritual path that women and men in the late twentieth century must follow to perfect their freedom and to find God. In developing this theme, special attention is given to the environmental crisis, which is forcing the human species to face its most fundamental problems.

Preliminary Reflections on the Meaning of Reverence for Life

A people's attitude toward life plays a critical role in shaping the way a society develops. This attitude is formed by a people's basic beliefs and values. It involves their intellectual and moral orientation in relation to the world. While it is true that the interconnected problems of poverty, violence, and environmental degradation can be solved only with the aid of major economic, technological, and political innovations, it is also the case that the most fundamental change needed is a transformation of human attitudes and values, leading people to adopt new ways of living and relating to each other, to nonhuman species, and to the larger world of nature. In the search for an understanding of the good life, this is an issue of primary importance.

These reflections suggest the vital significance of the religious dimension of life. The religious life develops in and through the quest for a liberating worldview and vision of the ideal. It matures with formation of a faith in a unified vision of those ethical and spiritual values that promise to guide the individual and society to well-being and fulfillment.

Before exploring the importance of reverence for life to a liberating religious and moral vision, it is necessary to clarify the meaning of the term *reverence*. It is not a word that is commonly used today. As one dictionary defines it, reverence is "a feeling or attitude of deep respect tinged with awe."[3] The awe and wonder associated with the feeling of reverence are often a response to what is perceived as sacred. Reverence for life may be defined, then, as an attitude of deep respect, involving feelings of awe and wonder and a sense of the sacred, before the mystery of life.

The idea of reverence for life is an ancient one. The history of the idea in the East and the West is intimately related to the story of the development of humanity's moral consciousness—its understanding of what constitutes good and evil, right and wrong, in human behavior. Reflecting on this history, three aspects of the idea stand out and help to clarify the meaning and contemporary significance of this ethical ideal.

First, in many different cultures and religions around the world, respect for life has led spiritual teachers to embrace Isaiah's principle: do not hurt or destroy. This is the first and basic ethical meaning of reverence for life, and it reflects an attitude of nonviolence. In the East one finds this teaching, for example, developed in ancient India by Hindus, Jains, and Buddhists. They spoke of *ahiṃsā*, a Sanskrit term that means "no harm." Gandhi rekindled interest in the principle of *ahiṃsā* in twentieth-century India by introducing it into the sphere of political action as a method of nonviolent political resistance. In the West the idea of *ahiṃsā* was, for example, taught by Socrates. In Plato's *Crito*, he defined evil as injuring another person, and he argued that it is never good to do evil. The first principle of the moral life, he explained, is the idea that "we ought not to retaliate or render evil for evil to anyone, whatever evil we may have suffered."[4] In a similar spirit, Jesus urged people not to return evil with evil but to free themselves of anger and to love even their enemies.

Second, as Jesus' emphasis on love (*agape*) suggests, the ethical implications of reverence for life go beyond *ahiṃsā*, noninjury, to embrace an active concern to help others. The Dalai Lama states the full ethical implications very clearly when he summarizes the meaning of Buddhism as follows: "All of Buddha's teachings can be expressed in two sentences. 'You must help others.' . . . 'If not, you should not harm others.' . . . Both teachings are based on the thought of love, compassion."[5]

Insofar as you are able, help others; and if you cannot help them, at least do not harm them. That is the basic moral ideal and guiding principle that flows from reverence for life.

We live in an imperfect world full of ambiguities, and situations arise, of course, in which the two principles of helping others and noninjury may come into conflict. In such circumstances a responsible person may choose to use coercive force and harm someone in order to help and prevent harm to others. Pacifism may not always be the best moral choice, but the ideal of respect for life establishes an attitude that is nonviolent in spirit. Those committed to it seek to find nonviolent methods of cooperative intelligence for resolving problems.

Third, when the Dalai Lama as a Buddhist teaches that one should not harm and should try to help others, he is not thinking only of human beings; he includes among those deserving moral consideration all living beings capable of experiencing pain and suffering. This, too, is an important aspect of the meaning of reverence for life. Furthermore, adoption of a biocentric worldview and ethic is essential if human civilization is to address the global problems that form the environmental and social crises that threaten its future.

The greatness of Western ethical thinking has been the way it has, over time, developed the values of respect for human life found in its religion and philosophy, contributing to powerful visions of moral democracy and to revolutionary social change. However, even though there are notable exceptions, Western religion and philosophy in general have been anthropocentrically oriented. Western thinking must now expand its respect for life from an appreciation of the inherent dignity of the human being to a recognition of the inherent worth of the entire community of life.

Among twentieth-century Western philosophers, Albert Schweitzer (1875–1965) stands out as the individual who laid the foundations for contemporary efforts to elevate and think through the meaning of reverence for all life as the supreme ethical value. The further reflections in this essay, which focus on the development and practice of reverence for life, are much influenced by Schweitzer's autobiography, *Out of My Life and Thought*, and his major work, *The Philosophy of Civilization*.[6]

Developing Reverence for Life

The attitude and practice of reverence for life begins to form with development of a deep respect for our own personal life. A human being naturally and instinctively values his or her own life and seeks to protect it and to grow unless raised in a social environment that is psychologically destructive. However, an attitude of reverence for life emerges only when this instinctive response to life is transformed into a fully conscious affirmation of life, involving a deliberate choice. This requires a deepening of self-knowledge and directly awakening to the mystery of life within ourselves.

Schweitzer argues that "the most immediate fact" of our self-awareness is the consciousness that "I am life which wills to live in the midst of life that wills to live."[7] This willing to live, furthermore, is not just a will to survive. We seek to avoid injury and pain and to achieve well-being and happiness. As Socrates put it, people do not just want life but the good life.[8]

If we search more deeply into our basic being as will-to-live we discover a further truth. To achieve this we must look behind all the masks we don for the world and penetrate beneath whatever we carry in the way of anger, self-doubt, fear, sadness, and loneliness. When we make this journey inward, looking deeply into ourselves and becoming consciously what we truly are, we find at the core of our being a wonderful free-spirited spark of life that is inherently good and beautiful. This is to discover what the Bible calls the heart, the deeper center of human nature where thinking, feeling, and willing are one. This deeper center exists in and has its being from the divine light that illuminates the darkness of existence. This wonderful, inherently valuable spark of life yearns for wholeness and relatedness, for freedom and fullness of life. Its presence is often revealed on the faces of little children.

To assert that our will-to-live is inherently good is not to deny that in all of us this spark of life is imprisoned to one degree or another by ignorance and self-centeredness. However, to become aware of this mysterious and wonderful will-to-live at the core of our being is itself a step toward freedom, for through this awareness one becomes intimate with the reality and truth of life itself. Through this practice one can encounter what Edward Hicks called the Inner Light.

Having achieved this new level of self-awareness, we must decide whether to affirm the will-to-live within ourselves and commit ourselves to its full development and perfection. If we fail to commit ourselves, we are unfaithful to our own true nature. We must choose. The attitude of reverence for life begins to take form in us when we affirm our own life and pursue in concrete ways the tasks of healing injuries suffered and of nurturing our growth inspired by a vision of our ideal possibilities.

As self-knowledge deepens, a person becomes keenly aware that he or she exists as a being interrelated with the community of life and with other sparks of life that yearn for fullness of life. Through the power of imagination human beings are able to identify with the quest for fullness of life in others. There are many social forces that work to limit the extent of this identification. They include racial prejudice, sex discrimination, ethnic hatred, religious exclusivism, and anthropocentrism. However, the more aware individuals are of the sacredness of the life in their own being and the more they experience the joy of growing and creating, the more powerful is the compulsion to identify with the suffering and struggle for well-being in other persons and creatures. Sympathy and compassion arise naturally from the heart, especially in a community that encourages these feelings and supports those who act on them.

This phenomenon reflects a simple truth recognized by teachers of universal love and compassion in all cultures and by our best democratic social ethics. This simple truth affirms that what we all share as human beings—our basic nature, our capacity to reason and to love, to weep and to experience joy—is more fundamental than all that divides and separates us in the way of gender, race, religion, ethnic origin, sexual orientation, or political affiliation, important as these differences are. In this insight into the intrinsic worth of all persons lies the foundation of respect for individual rights and the struggle for justice. Here one finds the only sure basis for enduring community in the midst of pluralism and for world peace. It is the essential truth taught by the Inner Light. It is the message of reverence for life.

Furthermore, in keeping with the spirit of Hicks's paintings, teachers like Schweitzer, Gandhi, and the Dalai Lama would, in the name of reverence for life, have us recognize that the life we share in common as humans links us spiritually as well as physically with all that lives and the great community of life that is planet Earth. In our time it is the astronauts' photographs of planet Earth—images of an illuminated, a beautiful, fragile sphere floating in dark space—that provide the most compelling contemporary symbol of community with all people and nature. This image of beauty and wholeness calls us back to our true nature and ideal possibilities. It invites affirmation of the meaning and goodness of life in nature and commitment to the ideal of respect and care for all life.

Albert Schweitzer has explained simply and clearly the moral definition of good and evil from the standpoint of reverence for life. For Schweitzer, "life, as such, is sacred."[9] His formulation of the supreme moral principle, therefore, extends Jesus' golden rule—"Do unto others as you would have them do unto you"—to all life. What is good, then, is "to preserve life, to promote life, to raise to its highest value life which is capable of development." Conversely, what is evil is "to destroy life, to injure life, to repress life which is capable of development."[10] Schweitzer's definition of good and evil is a refinement of the principle: Insofar as you are able, help others; and if you cannot help, at least do not harm others.

The Scope of Reverence for Life

In *The Philosophy of Civilization*, Schweitzer extends the attitude of reverence for life beyond plants and animals to "every existing thing." For example, he speaks of there being a will-to-live "in the flowering tree, in the strange forms of the medusa [a jellyfish], in the blade of grass, in the crystal; everywhere it strives to reach the perfection with which it is endowed. In everything that exists there is at work an imaginative force, which is determined by ideals."[11] "True knowledge consists in being gripped by the secret that everything around us is will-to-live," writes Schweitzer.[12]

These words remind one of Alfred North Whitehead's process philosophy, which rejected the Cartesian and Newtonian view of nature as essentially a machine made up of dead matter in motion. Whitehead argued that the concept of life is essential to an understanding of nature at all levels. He explains that "neither physical nature nor life can be understood unless we fuse them together as essential factors in the composition of 'really real' things whose interconnections and individual characters constitute the universe."[13] Drawing on the new physics and biology, Whitehead viewed the whole universe and everything in it as active, alive. He was not a panpsychist who believed that conscious conceptual thought exists in all things, but he did believe that to one degree or another activity, aim, feeling, and enjoyment of intrinsic value exist everywhere. When reverence for life is coupled with a process metaphysics such as Whitehead's, the principles of *ahiṃsā* and promotion of life embrace human relations with all that is.

The life-oriented philosophies of Whitehead and Schweitzer remind one also of the way the nineteenth-century European Hasidic masters used the Kabbalistic myth of the holy sparks. They taught that in all things, including stones, tools, plants, and animals, dwell divine sparks. It is the special task of humankind to respect everything in the natural world and to use and interact with all things in such a way as to liberate these divine sparks. When this occurs, all living becomes sacred, holy, and the natural world is freed to be reunited with God. Martin Buber was much influenced by Hasidism, and his I–Thou philosophy is an attempt to clarify the kind of relationship that would liberate the world and overcome the separation of the sacred and the secular.[14]

A Zen story told by the Japanese philosopher Keiji Nishitani also illustrates the point. After mopping the floor of the kitchen, a young monk threw a bucket of water out onto a cement roadway. Pointing to a nearby tree, his teacher scolded him for his thoughtlessness with the command: "Let the water live!"

Today the new ecology and theory of the evolution of the universe can only reinforce and deepen the attitude of reverence for life. We understand with new insight the interdependence of our self with the larger community of life and the universe as a whole. There is no sharp line of demarcation between one person and another, between humanity and nature, between the community of life on earth and the fifteen-billion-year process of cosmogenesis that is the universe. Each person is a manifestation of the cosmic energies that have created and continue to create the universe. Each is a unique expression of the totality. We are in the totality and it is in us. The universe is a dynamic process involving diversity in the midst of unity. This reality is what some American Indians appropriately call the Great Holy Mystery. To imagine a salvation of my self or a fulfillment of my group separate and isolated from the well-being of the larger community of being is, in the final analysis, an illusion. Through the environmental crisis the earth is teaching humanity this anew, awakening women and men to what Joanna Macy has called "the ecological self" and to what Aldo Leopold described as "the ecological conscience."[15]

An international committee of thirty-four leading scientists headed by Carl Sagan recently issued a statement in which they wrote that "many of us have had profound experiences of awe and reverence before the universe." They urged people everywhere to regard "our planetary home" as "sacred," and "to safeguard and to cherish the environment."[16]

The point is that the entire earth is of intrinsic value quite apart from its utilitarian value to people, and people are interconnected with it physically and spiritually. If we would fully understand these insights, we must act on them and allow the spirit of respect and care to pervade our daily lives and all our interactions with the world around us. This is, of course, a demanding discipline that requires that we slow down, smile more, practice mindfulness, and dwell wholeheartedly in the present. Both we and our world will benefit from the effort. It will erode egoism and awaken us to the beauty all about. Where there had been separation, intimacy and community will blossom. We will be rewarded with a sustaining sense of belonging to the whole and a feeling of enduring peace.

A Path to Freedom, Faith, and God

In this connection it is instructive to reflect on the ways in which the practice of reverence for life is a path to inner freedom and to religious faith and relationship with God. Regarding the achievement of inner freedom, the ideal of reverence for life involves a commitment to be faithful to one's inmost self, to one's heart and its ideals, and this is the beginning of freedom from the world—that is, from a blind attachment to external things that enslaves the human spirit. On the path of reverence for life, a person begins to realize that the key to fulfillment and meaning is not what I have but what I am. What I am is shaped by what attitudes and values I choose to adopt in relation to myself and in my interactions with the world. Human beings frequently have no control over what the world does to them, but they do have the potential to determine their own attitude toward the world. Some philosophers have described this distinctively human capacity as the last freedom, which no one can take away from a human being. In this freedom lies the unshakeable foundation of human dignity. In this regard, it is worth recalling that Martin Luther King Jr. often fasted and prayed when he was imprisoned during his leadership of the civil rights movement, in an effort to purge his heart of hatred for his captors so that he would not get caught up in the very racial animosity that he was fighting.

Given the fears and desires that possess human beings, it is difficult to achieve this last and most essential freedom. However, the first step into it involves getting in touch with our heart and choosing to adopt an attitude of reverence for life in response to the mystery and wonder of our own will to full life. It deepens as the attitude of reverence for life is extended outward in response to the needs of others and the community.

The way of reverence for life involves a faith that is religious in nature. A healthy life-affirming religious faith involves a trust in the enduring meaning and value of life in spite of all the suffering and inexplicable evil that are encountered in existence. At the core of such a faith is a great Yes to life that wells up out of the depths of our being, possessing our minds and hearts. The experience of being grasped by the mystery, beauty, and inherent value of life is an encounter with the sacred. It is a religious experience and the awakening of faith. This faith inspires a person to seek fulfillment in and through spiritual growth guided by the ethics of reverence for all life.

Furthermore, to be possessed by a faith in life is to experience faith in God. God is the unfathomable primal whole and the source of life and of the truth, beauty, and goodness that fulfill life. The cosmic process is within God, and God is within the cosmos as the ultimate power of life.

One finds and experiences God in and through the encounter with the mystery of life and the practice of reverence for life. I have in mind a world-affirming ethical mysticism. In our time, God is not to be found by turning away from the world and seeking for the divine outside the world. Everything depends on the quality of our relationship with the world. One finds God by purifying one's heart and mind of delusion and egoism and by entering with one's whole being into relationships of respect and caring with the life all around us. It is in the act of loving that human beings are able to experience and know most deeply the God who is love. The light of divine meaning and goodness pervade human life just insofar as people wholeheartedly embody reverence for life in their daily existence. This is as true in times of great adversity as it is in times of relative peace. This faith is especially important today for those women and men who search for the courage to face and adjust to the grim truth regarding the current world situation.

Practicing Reverence for Life in Nature

At this juncture it is necessary to consider a very difficult and complex issue that cannot be avoided in any discussion of reverence for life as an ethical ideal. We live in an ever-changing universe, where all life sustains itself by consuming other life, and death follows birth. Last June my wife and I watched in dismay as a kestrel carried off a young tree swallow while its parents screeched in protest. I cut trees in the process of managing an apple orchard and landscaping, and I eat fish and fowl as well as a variety of grains and vegetables. It is not possible to live, grow, and create without killing other life and destroying the habitat of other creatures.

If the harsher realities of life in nature shock our finer sensibilities, we might ponder Gary Snyder's comment that "It is either 'nature red in tooth and claw' or a sacrament. Take your pick."[17] However, even if the consuming and being consumed that goes on in nature is somehow participation in a sacred mystery that is a form of communion, this still leaves us with the question: What does it mean to live by the ideal of reverence for life in this world in our time?

Most fundamentally, it means to take full responsibility for how we live. Being responsible means being aware of our interdependence with the larger community of life and being informed about the state of the planet, the nature of our bioregion, and the harm and suffering that our actions may cause other people and nature. It also means building strong human communities energized by the spirit of participatory democracy and cooperative problem solving. It involves supporting local, national, and international strategies for sustainable development in an effort to ensure that future generations inherit a world that is as beautiful and supportive of life as the one present generations inherited. As families, institutions, and whole communities begin to live and operate in sustainable ways, the social environment will itself begin to shape human attitudes and values in positive new directions.

Sustainable development requires pollution prevention, minimizing waste, recycling, energy efficiency, reducing consumption, simplifying lifestyles, conserving biodiversity, protecting natural beauty, developing environmentally benign technologies, integrating environmental concerns into all economic decisions, stabilizing human populations, and working to overcome poverty wherever it exists.[18] In the face of the worsening environmental crisis, the need for major changes along these lines is urgent. Continuation of current government and economic policies will have disastrous consequences.

Practicing reverence for life also means recognizing that animals, trees, rivers, and ecosystems possess intrinsic value and are not merely things to be used, even though humans must use them to live. If we do use other life forms and resources with awareness, we will act with humility and gratitude. We will recognize that since nonhuman beings and also ecosystems possess intrinsic value, they are in a real sense ends in themselves and as such are worthy of respect and care for their own sakes. It is for this reason that a number of philosophers, theologians, and activists argue that nature possesses rights and humanity has an obligation to treat animals, plants, and biotic communities with justice by respecting their rights.

The notion that humanity should affirm the rights of dolphins, maple trees, and the biosphere as a whole and expand its concept of justice accordingly is a controversial idea today. Rights theory offers one way of trying to develop the implications of reverence for life and to clarify the nature of humanity's moral responsibilities in relation to nature. Some people prefer simply to talk about the feeling of deep respect for life and the sense of moral responsibility it awakens without using the language of rights. This is a reasonable position, especially in the light of the difficulties one has in sorting out just what the rights of nonhuman creatures are and what they mean for human behavior. Some philosophers opt to use rights language in connection with only a part of nonhuman nature, limiting its application, for example, solely to animals.

There are some advantages in introducing the concepts of justice and rights into the discussion of the ethics of reverence for life.[19] It sharpens the moral issue and the claim of the nonhuman world for ethical consideration. In its most fundamental form justice has long been viewed as giving each member of the community his or her fair share of the necessary and good things in life. The theory of human rights seeks to define what a person needs and what he or she is entitled to in order to be able to enjoy life, liberty, and the pursuit of happiness. By extending the concepts of rights and justice to animals, plants, and nature, it affirms the continuity of humanity and nature. It provides a way of focusing the discussion of human moral responsibilities. It also facilitates the construction of effective legislation and the protection of nature through the judicial system as, for example, in the case of the U.S. Endangered Species Act.

An adequate theory of humanity's moral responsibilities for, or the rights of, nature must address at least three distinct but related areas of concern. First, it requires guidelines for the treatment of individual creatures, which, for example, has been the concern of animal rights advocates. In this regard, groups like the Humane Society have developed very useful guidelines for the treatment of domestic animals; and James Nash's treatise on environmental ethics, *Loving Nature,* contains an instructive "Bill of Biotic Rights" that tries to define human responsibilities in relation to wild creatures.[20]

Second, a complete vision of human obligations also involves inquiry into humanity's responsibilities for species. The objective is to address the kind of question that arises when humans threaten to destroy the habitat on which an entire species is dependent or seek to eliminate a deadly virus like HIV-1. In this regard, it is noteworthy that the UN World Charter for Nature, which was adopted by the UN General Assembly in 1982, states that "Every form of life is unique, warranting respect regardless of its worth to man, and to accord other organisms such recognition, man must be guided by a moral code of action."

Third, a theory of the rights of nature involves what Aldo Leopold called a land ethic that respects and protects the well-being of biotic communities as wholes, that is, the ecosystems upon which all life depends.

In New England, for example, the perspective provided by a land ethic is urgently needed to guide governments, communities, and corporations in saving and restoring the great northern forests that extend from the Adirondacks to the Maine Woods.

The relation of human rights to the needs and interests of nonhuman nature raises a difficult set of questions. In this regard it can be said, first of all, that since we live in a sacramental universe and life as such is sacred, cruelty, unnecessary killing, and careless harming and destruction of anything are morally irresponsible. Humans have no right to act in this way.[21] In this regard, animal rights activists and conservationists frequently identify important moral issues from the cruelties of factory farming to the protection of wetlands and the ozone layer.

Second, the human species finds itself existing in a world where killing and the use of resources is necessary for living, and, therefore, human beings are justified within strict limits in using other life forms and in destroying some habitats of other species in order to survive and to improve the quality of human life. It is important to keep in mind that the rights of human beings are far more extensive and complex than those of other creatures, reflecting humanity's unique capacities. In addition, the rights of nonhuman nature are not absolute, which is, of course, also true of human rights. For example, a criminal may lose his right to liberty, and whatever rights a fish may have to life and liberty may, under certain circumstances, be subordinated to a human being's need for food.

When confronted with the dilemma of when, where, and how to sacrifice life in order to protect, sustain, and save other life, humans must weigh and balance all the competing interests and values in a situation, considering carefully both the long- and short-term consequences of alternative courses of action. If as they work to improve the quality of human life, they act in the spirit of *ahiṃsā,* they will endeavor to minimize the pain, killing, and damage they inflict on other life forms and nature. They will ensure that the well-being of the community of life as a whole is not threatened. As a general rule, where possible, they should seek to help with the healing of wild nature when human activity has caused damage. The old gold miner in the film *The Treasure of the Sierra Madre* had the right idea when he insisted on restoring the mountain that he had dug up and then thanked the mountain for its gifts when he departed.

Among those pursuing ecotheology and environmental ethics today there is a group that has tried to build on and develop the reverence for life philosophy of Schweitzer. Their analysis of the issues and moral visions are an especially important contribution to contemporary environmental philosophy. However, some of these thinkers have adopted one idea against which Schweitzer warned, and their thinking on this subject is problematical.[22] In their search for criteria with which to make moral judgments, they employ the idea of a hierarchy of beings in nature, with humans at the top, and they argue that this hierarchy reflects varying degrees of intrinsic value in different life forms. All beings have intrinsic value, but some have more and some less. Roughly speaking, the intrinsic value of a creature is to be measured, according to this theory, by the degree to which its experience approaches the qualities of human experience.

There is no doubt that nature is characterized by diversity and that as a practical matter human beings must make some general judgments about the relative priority to be given to the interests of different life forms. However, our determinations in this matter are based on a human perspective, and they should not be taken as judgments about the intrinsic value of other beings. As Schweitzer comments: "Who among us knows what significance any kind of life has in itself, and as a part of the universe?"[23]

In addition to Schweitzer's question, both religious and moral objections can be raised against the idea of gradations in the intrinsic value of different life forms. First of all, from a religious perspective, it may be argued that intrinsic value is not something that exists in measurable quantities. The idea of intrinsic value is derived from a sense of the sacred. Things have intrinsic value because they are members of the great community of life and the divine mystery is at work in them. In Schweitzer's philosophy what gives things intrinsic value is the presence of life, and he writes that "life, as such, is sacred." Life is not more sacred in this being than in that. As the medieval mystic and theologian Meister Eckhart pointed out, "God loves all creatures equally and fills them with his being. And we should lovingly meet all creatures in the same way."[24] From this point of view, each unique life form is perfect, whole, and complete just as it is, without qualification.[25]

The intrinsic value in things is to be identified with what Martin Buber called the "thou" in all things. Making gradations in value is I–it talk, and it is not possible in the realm of I–Thou. The thou in things cannot be turned into an object for analysis and quantified. It is the presence of this mysterious, elusive, sacred thou in all living beings that constitutes their intrinsic value.

There are moral problems, too, with using the idea of a hierarchy of beings based on differing amounts of intrinsic value. There is the danger that this notion may be used to support the kind of anthropocentric attitude that has led to hubris in humans in their relations with nature, causing them to exceed the limits of what is just and what nature will tolerate. Furthermore, hierarchical thinking has haunted human culture with destructive distinctions regarding the relative worth of people based on gender, race, religion, ethnic origin, and class. The feminist critique of patriarchal social structures has been especially effective in exposing the way hierarchical and dualistic modes of thought have been used to justify the subjugation and domination of one group by another. As human history illustrates, it is very easy for human thought to move from the idea of the superior worth of the human species to the idea of the superior worth of the people of one particular gender, race, or religion.[26] The call of ecofeminists for creation of inclusive communities that emphasize mutuality, sharing and cooperation rather than relationships of domination and control is an expression of the values of respect for life, especially when the call includes a concern to free nature of unjust unsustainable exploitation. Among the ethical and religious thinkers who employ the idea of gradations of intrinsic value, there are some who recognize and seek to address the social and environmental problems that have accompanied hierarchical thinking, but they fail to recognize the degree to which this notion is one of the roots of the trouble.

In the light of these concerns, it would be best if the ethics of reverence for life abandoned all discussion of degrees of intrinsic value in connection with judgments that involve the sacrifice of one life to protect or promote the life of another. It is sufficient to focus instead on the specifics of the situation at hand and on the concrete concerns of intelligent compassion, such as minimizing pain and suffering, preserving the welfare of the whole, and improving the quality of human life. There is no need to introduce a discussion of levels of intrinsic value. It is an unnecessary rationalization.

It is important at this juncture to emphasize that the ethical ideal of reverence for life should become the first principle of a world ethic of sustainable development as well as a guide in individual living. Given the growing global interdependence of all peoples ecologically, economically, politically, and culturally, only the creation of a world community united by shared values and committed to justice, peace, and sustainable living can ensure the flourishing of human civilization on earth in the twenty-first century. This observation is not meant to deny the value of cultural diversity or that there are many different cultural pathways to a unifying vision of shared values. However, without unity in the midst of diversity, the future history of the human species will disintegrate into self-destructive ethnic, religious, racial, and national conflicts. The essential unity needed is a shared faith in an ecological and democratic world ethic that is rooted in reverence for life.[27]

All of the world religions have a special responsibility to cooperate together and with other morally concerned organizations in creating and instilling this new world ethic of respect and care for the whole community of life. A fine example of such cooperation is the new National Religious Partnership for the Environment in the United States, which has brought together leaders from the Roman Catholic, Orthodox, mainline Protestant, evangelical, and historic Black churches and from the major branches of the Jewish religious community to cooperate in the areas of environmental education and protection.[28] However, an enormous amount of work remains to be done by the religions to bring their institutional organizations into full support of the ethics of reverence for life and sustainable development.

That life in nature means impermanence, suffering, and death cannot but disturb us with sadness. That human actions greatly intensify the suffering of the world and now pose a threat to the planet's basic life support systems can only deepen this sadness and add to it a sense of great human moral failure. A mystery surrounds the existence of evil and suffering in nature that is beyond our understanding. Nevertheless, there is a light that shines in the midst of the darkness. If human beings search deeply enough, in their hearts they know the sacredness of life and the goodness of existence. In the human heart there is also a passion for a freedom that can find fulfillment only in community with all life. Here lies the wisdom of Isaiah's vision, Edward Hicks's paintings, and the images of planet Earth. If we are faithful to this inner light in a spirit of humility, gratitude, and care, there awaits us a sense of meaning and peace that neither evil nor death can take away. It becomes, then, the part of wisdom to face the darker mysteries of existence with a faith in life and a quiet determination to do what we can to preserve and promote it. In conclusion, our best hope for creating community and healing the earth lies in understanding ourselves as citizens of the universe and in putting our trust in the wisdom of reverence for life.

Notes

1. See Eleanore Price Mather, *Edward Hicks: A Peaceable Kingdom* (Princeton, N.J.: The Pyne Press, 1973), Introduction. In some Hicks paintings the presence of Christ is explicitly symbolized by the use of a grapevine growing from the tree that represents "the stump of Jesse" and the House of David (Isa. 11:1). See, for example, Hicks's painting *The Peaceable Kingdom of the Branch*. It is also noteworthy that many of Hicks's paintings include William Penn, the seventeenth-century Quaker founder of Pennsylvania, signing a legendary peace

treaty with Indians. This scene was for the artist a symbol of his faith in America and hope that in this land of liberty men and women inspired by the Inner Light might follow Penn's example and work to realize the kingdom of peace on earth. See also Alice Ford, *Edward Hicks: His Life and Art* (New York: Abbeville Press, 1985).

2. The theme of Isaiah 11:6–9 is repeated in Isaiah 65:25.

3. *Random House Dictionary*, 2d ed.

4. Plato, *Crito*, in *The Works of Plato*, trans. B. Jowett, ed. Irwin Edman (New York: Modern Library, 1956), 99.

5. His Holiness the Dalai Lama, in *The Dalai Lama, A Policy of Kindness*, ed. Sidney Piburn (Ithaca, N.Y.: Snow Lion, 1990), 88.

6. Albert Schweitzer, *Out of My Mind and Thought: An Autobiography*, trans. C. T. Campion (New York: Henry Holt and Company, 1933, 1949), chap. 13 and Epilogue; Albert Schweitzer, *The Philosophy of Civilization*, trans. C. T. Campion (Buffalo, N.Y.: Prometheus Books, 1987), pt. 2, Preface, chaps. 6, 22–27.

7. Schweitzer, *Out of My Life and Thought*, 75.

8. Socrates, *Crito*, 98.

9. Schweitzer, *Out of My Life*, 158–59.

10. Ibid., 158. See also Schweitzer, *The Philosophy of Civilization*, 309–10.

11. Schweitzer, *The Philosophy of Civilization*, 282; see also 308–10, 318.

12. Ibid., 325. See also 308, 332–33.

13. Alfred North Whitehead, *Modes of Thought* (New York: Capricorn Books, 1958), 205.

14. Martin Buber, *Hasidism and Modern Man* (New York: Harper & Row, 1958), 28–37.

15. Joanna Macy, *World as Lover, World as Self* (Berkeley, Calif.: Parallax Press, 1991), chap 7; and Aldo Leopold, *A Sand County Almanac* (New York: Ballantine Books, 1966), 243–46.

16. *An Open Letter to the Religious Community from the Scientific Community*, January 1990 (available through the National Religious Partnership, 1047 Amsterdam Avenue, New York, NY 10025).

17. Gary Snyder, "Rediscovering Turtle Island," John Hamilton Fulton Memorial Lecture, Middlebury College, Middlebury, Vermont, October 26, 1993.

18. For an example of a national policy for sustainable development, see *Choosing a Sustainable Future: The Report of the National Commission on the Environment* (Washington, D.C.: Island Press, 1993).

19. James Nash provides a very instructive discussion of these issues in his *Loving Nature: Ecological Integrity and Christian Responsibility* (Nashville, Tenn.: Abingdon Press, in cooperation with the Churches' Center for Theology and Public Policy [Washington, DC], 1991), chaps. 6–7.

20. Jay McDaniel, *Of God and Pelicans: A Theology of Reverence for Life* (Louisville, Ky.: Westminster/John Knox Press, 1989), 70; James Nash, *Loving Nature*, 186–88.

21. See "Elements of a World Ethic for Living Sustainably," in *Caring for the Earth: A Strategy for Sustainable Living* (Gland, Switzerland: IUCN–The World Conservation Union, UNEP–United Nations Environment Programme, and WWF–World Wide Fund for Nature, 1991; available through Island Press, Covelo, Calif.) 14. The American poet, farmer, and environmental philosopher Wendell Berry recently commented in response to questions about the protection of wildlife and natural resources: "Nobody has a right to destroy anything, and everybody has an obligation to defend as much as he or she possibly can." (Jordan Fisher-Smith, "Field Observations: An Interview with Wendell Berry," *Orion: People and Nature* 12 [autumn, 1993]: 56.

22. See, for example, John Cobb and Herman Daly, *The Common Good: Redirecting the Economy toward Community, the Environment, and a Sustainable Future* (Boston: Beacon

Press, 1989), 378–79; Jay McDaniel, *Of God and Pelicans*, 60–84, and *Earth, Sky Gods and Mortals: Developing an Ecological Spirituality* (Mystic, Conn.: Twenty-Third Publication, 1990), 66–67, 91–82; and James Nash, *Loving Nature*, 149–50, 181–83. Even though I disagree with these authors regarding the issue of degrees of intrinsic value, I find their larger moral and religious visions persuasive in many respects.

23. Schweitzer, *Out of My Life and Thought*, 233. Schweitzer adds: "To undertake to lay down universally valid distinctions of value between different kinds of life will end in judging them by the greater or lesser distance at which they seem to stand from us human beings—as we ourselves judge. But that is a purely subjective criterion."

24. See Matthew Fox, *Breakthrough: Meister Eckhart's Creation Spirituality in New Translation* (Garden City, N.Y.: Doubleday, 1980), 92.

25. Some ecotheologians argue both that God loves all creatures equally and that there are degrees of intrinsic value in nature, but this appears to this author to involve a contradiction.

26. See, for example, Rosemary R. Ruether, *Gaia and God* (San Francisco: HarperSanFrancisco, 1992).

27. The beginnings of such an ethic have already been sketched in the World Conservation Strategy published by IUCN, the International Union for the Conservation of Nature, based in Switzerland. See *Caring for the Earth: A Strategy for Sustainable Living*. IUCN is the largest environmental organization in the world. Its members include 99 governments and 575 nongovernmental organizations (NGOs), and it cooperates closely with the United Nations. What is especially significant about the World Conservation Strategy is its recognition that ethical vision and commitment are essential to any global plan for achieving sustainable development. *Caring for the Earth* asserts that the first principle of a world ethic should be "respect and care for the community of life" (pp. 8–9). It is this principle that should govern the development of the new technology and the new economics as well as individual life-styles. The ongoing efforts to create a UN Earth Charter are part of an international effort to build on the work of IUCN and other groups. See also Hans Küng, *Global Responsibility: In Search of a New World Ethic* (New York: Crossroad, 1991).

28. The National Religious Partnership for the Environment, 1047 Amsterdam Avenue, New York, N.Y. 10025.

Part II | OLD PATHS, NEW GROUND

How wonderful, O Lord, are the works of your hands!
The heavens declare Your glory,
the arch of sky displays Your handiwork.
In Your love You have given us power
to behold the beauty of Your world
robed in all its splendor.
The sun and the stars, the valleys and hills,
the rivers and lakes all disclose Your presence.
The roaring breakers of the sea tell of your awesome might,
The beasts of the field and the birds of the air
bespeak Your wondrous will.
In Your goodness You have made us able to hear
the music of the world. The voices of loved ones
reveal to us that You are in our midst.
A divine voice sings through all creation.

—Jewish Prayer

Rabbi Everett Gendler is rabbi of Temple Emanuel in Lowell, Massachusetts, and is Jewish chaplain and teaches at Phillips Academy, Andover, Massachusetts. The basic theme of his chapter is that in following old paths and traditions we need also to discover new liturgical grounds to help us reorient ourselves in God's creation. In other words, "join the chorus, recapture the rhythms."

Rabbi Gendler challenges the occasional tendency to assign ultimate responsibility for the ecological destruction wrought in this century to the Creator, thereby turning him/her into a sort of "cosmic culprit." However, the psalms of praise of that creation, as well as the Noah Covenant with all of nature's many entities, suggest that Lynn White's earlier assertion that the Jewish/Christian tradition is fundamentally responsible for the contemporary ecological crisis was perhaps overhasty.

Indeed, Jewish tradition spells out a number of classical ecological principles in its understanding of how humans are to behave toward nature. The good life is not material consumption or even proud success and recognition but humbly finding our way to "an environmentally responsible and religiously appreciative life."

Such a life seems far from the modern, all too common radical alienation from nature in which we seem cut off and separated from it, bloodlessly abstracted out of the only vital context there is for life. We need, then, to find new liturgical ways to reintegrate our lives within the encompassing vital rhythms of nature. "Thus do we celebrate, and come to cherish all the more, the fresh gifts of each season. And from the cherishing must surely come our ever deepening caring for this planet."

Calvin DeWitt, professor of environmental science at the University of Wisconsin–Madison and director of the Au Sable Institute of Environmental Studies in Michigan, writes on ecology both as a working scientist and a biblically centered evangelical Christian. Although the ecological crisis we face is new ground for Christians in that it has emerged only recently as the significant and ominous issue it is, Professor DeWitt believes there are both scientific and scriptural grounds for Christians today to be actively engaged in dealing with it.

First of all, science has shown us that "the present environmental state of the world constitutes the most serious threat to the biosphere since the origin of life on earth." As an evangelical Christian, Professor DeWitt believes that this violent attack on and degradation of nature is at root a spiritual issue because, as he puts it, "how one relates to Creation reflects how one relates to and honors the Creator."

The Bible reveals that God is a just, righteous, and loving creator of the entirety of an ordered (Torah) and good creation, including humans, of course. The "good life," then, is to glorify God in our lives, to live according to God's will and in harmony with his creation. Such a good life, Dr. DeWitt argues, entails following a number of ecological principles spelled out in the Bible. The ethical standards of the Bible, then, provide the basis for actually living differently and thus behaving differently toward God's creation. Living rightly, he tells us, must be added to our scientific understanding of how nature works if we are to come to grips with the shocking degradation of it that we are witnessing all about us.

Jay McDaniel is professor of religious studies and director of the Steele Center for the Study of Religion and Philosophy at Hendrix College, Conway, Arkansas.

He begins chapter 6 in a dramatic fashion by directly and starkly pointing to what he thinks is the prevalent religion of our day, what he calls economism or growthism. The god of this consumer society religion is endless progress and growth; "its priests are economists; its missionaries are advertisers; and its church is the mall. . . . salvation comes through shopping alone." It is this religion of economism, he thinks, that has led to our present environmental degradation and crisis.

Such a crisis, he tells us, cannot be resolved by technology or social policy alone, for it involves "character." In other words, it is a crisis founded on who we are and how we see life and our place within it. It is a religious crisis, then, and calls for religious and spiritual response.

Professor McDaniel then highlights three ideas that he believes Christians should understand in our present situation. First of all, the ecological problem is not one among a number of equally significant issues but rather is the context for understanding all other issues. Second, the "good life" is (or ought to be) neither material accumulation in the present life nor a better life after this one but rather a shalom-filled journey into wholeness. Finally, he tells us, God is ultimately mystery, not an entity external to creation but, as Godself, the sacred whole of the cosmos.

"Green grace," McDaniel argues, is the experience in which nature in its interdependence enriches us and helps to re-create us. "Red grace," on the other hand, is the experience of Eucharist, through which we identify with the suffering of others (including the wider community of nature) in order to transform that suffering through love.

Finally, we are asked by Dr. McDaniel to take seriously the idea that there is a divine purpose at work in the sacred whole of creation. Since, as he puts it, "the sacred universe itself is the very body of God," then we are called to enter into communion with that body, thereby helping to restore the original hopes of the Creation. Thus, we are called by God to participate creatively in healing and helping to bring nature to a more harmonious and balanced wholeness.

In chapter 7, Albert Fritsch writes on healing the earth in the context of a resurrection-centered theology. Father Fritsch is a Jesuit priest, an environmental scientist, and director of Appalachia–Science in the Public Interest in Kentucky.

In contrasting creation-centered spirituality with that which is resurrection-centered, Father Fritsch suggests the developing breadth of Catholic thinking on ecological issues. He centers the Catholic Christian's concern for planet Earth and its well-being on the Covenant with Noah in the book of Genesis and in the Eucharist Thanksgiving in which we accept the mission of Jesus to world and planet.

There are three stages of human growth vis-à-vis God's creation. First of all, God is recognized to be acting there in renewing and making it anew. That is, we first come to see God's creative activity in sustaining all of life. But, second, we are invited to act to save and heal the earth we have harmed by acknowledging our own sin and involvement in its degradation and by redeeming ourselves in the blood of Christ. Finally, we see, through the Holy Spirit, that we can come to live the good life of love and caring for all creation.

Lasting and effective environmental activism flows less from an enthusiasm for creation (creation theology) than from a recognition that we have damaged the seamless web of nature and need to restore it (resurrection theology). "In healing earth," he tells us, "earth heals us; in making earth new, we are renewed in and through earth."

Join the Chorus, Recapture the Rhythms

God

All too often, God is viewed as the problem, the cause of our environmental predicament, and hence deliberately excluded from books like this. There is a temptation to speculate about the reasons for this state of affairs, and to this temptation I shall only briefly succumb. Partly, I think, it is an overreaction to or a distortion of the thrust of the searchingly appropriate questions raised a quarter of a century ago by Lynn White Jr. All too easily we assume that a God who transcends nature thereby depreciates it; this is not necessarily the case, as I shall soon try to establish. But partly, I fear, it reflects a certain prevalent despair about our human situation today. Disheartened by the terrible destruction we have wrought by wars through this century and dismayed by the horrifying specter of genocide this century— first of the Armenians, then of the Jews, and now elsewhere as well—we understandably question the ultimate aims and purpose of creation and attempt to assign responsibility for the problematic if not downright miserable state of affairs. Quickly we move from awareness of our own failures to the attempt to assign ultimate responsibility for it all, and so we look toward the purported Creator, tending in this context to view God as cosmic culprit. This clearly enters the realm of theodicy, and given constraints of time we must largely resist further exploration of this issue, relevant though it may be.

There is, however, one characteristic of creation as such that does need to be recognized in clarifying the relation of environment and Creator. In the words of Alon Goshen-Gottschein:

. . . before the world's creation . . . God was the only reality and therefore all reality was One. The creation of a reality that could view itself as being separate from God—though it might not ultimately be so—signaled a transformation in reality as it had been up to that point. It was no longer unity; multiplicity had been introduced as a mode of existence. . . . Thus, if we understand God's unity as the ultimate unity of all existence, we must view creation as that process through which fragmentation and multiplicity enter a hitherto unified reality. . . . creation itself introduced duality and thus multiplicity into reality. Multiplicity is the hallmark of creation. ("Creation," in *Contemporary Jewish Religious Thought*, ed. A. A. Cohen and P. Mendes-Flohr [New York: Scribner, 1987] p. 11)

It is this multiplicity that yields both variety *and* imperfection. Here lies the source of that haunting problem, theodicy: injustice, and imperfection in a divine creation. For were there perfection, would not all again be God? And so this imperfect creation that we inherit and enjoy might again become nonexistent.

But *why* creation? "The quality of loving kindness (chesed) is the basis of all creation. It is God's steadfast love that brought this world into being, and it is God's steadfast love that maintains it." (Rabbi David S. Shapiro)

While there are more complex answers to the question, this sense of creation as a divine gift is found in nearly every alternative formulation. One might also dare to say that God so loved this world that s/he created it in all its fullness!

As for the quality of this gift, again the words of Shapiro: "The goodness of God . . . was objectified in the created world. This world is such that humans dare not degrade any of its parts. 'The One Who has created them praises them, who dares deprecate them? Their Creator lauds them, who can find fault with them?' (Genesis Rabbah 12:1)."

The Environment

This brings us immediately, if with undue haste, to what we call environment, the created order that surrounds us. Here the central question is the nature, the characteristics of this environment.

It is widely assumed that since biblical thought is not animistic, it therefore necessarily reduces nature to an inanimate state devoid of any sentience, any degree of feeling. "By destroying pagan animism, Christianity made it possible to exploit nature in a mood of indifference to the feelings of natural objects" (Lynn White Jr., "The Historical Roots of Our Ecological Crisis," *Science* 155 [March 1967]: 1203–7).

The first is the use of the term *brit* (covenant) in early Genesis. The first use of the term is in relation to all the life of our planet, not just human beings. Briefly cited in "The Earth's Covenant," let me here reproduce it with italics and somewhat expanded discussion.

And God said to Noah and to his sons with him, "I now establish My *covenant* with you and your offspring to come, and with every living thing that is with you—birds, cattle, and every wild beast as well—all that have come out of the ark, every living thing on earth. I will maintain My *covenant* with you: never again shall all flesh be cut off by the waters of a flood, and never again shall there be a flood to destroy the earth."

God further said, "This is the *sign* that I set for the *covenant* between Me and you, and every living creature with you, for all ages to come. I have set My bow in the clouds, and it shall serve as a *sign* of the *covenant* between Me and the earth. When I bring clouds over the earth, and the bow appears in the clouds, I will remember My *covenant* between Me and you and every living creature among all flesh, so that the waters shall never again become a flood to destroy all flesh. When the bow is in the clouds, I will see it and remember the everlasting *covenant* between God and all living creatures, all flesh that is on earth. That, "God said to Noah," shall be the *sign* of the *covenant* that I have established between Me and all flesh that is on earth." (Gen. 9:8–13)

Upon first consideration, the rationalist reader is likely to dismiss the wording as merely a figure of speech, a stylistic conceit perhaps. Yet the sevenfold repetition of *brit* and the threefold employment of *ot* (sign) in this passage forbid such easy dismissal of the implications. Seven and three, after all, are both sacred and efficacious numbers for the biblical outlook. (Cf. "A three-fold cord is not quickly broken" [Eccles. 4:12].) Both the repetitions and their specifics fairly insist that the notion of divine covenant in relation to the earth and its life be taken with utmost seriousness. While the covenantal references do in four instances specify human beings, in those same four instances the other living creatures are included as well. Two others refer generally to all living creatures, while the seventh speaks only of God's covenant with earth.

To take seriously God's covenant with other living creatures as well as with the earth itself raises a question at once disconcerting and exciting. Insofar as covenant is a term of reciprocity, involving an exchange of responsibilities and duties, what does this imply about the ontological status of the earth and its living creatures? Is the earth itself in some significant sense a living being? One of the greatest biblical scholars of the twentieth century, the late Johannes Pedersen, so argued in his magisterial work, *Israel: Its Life and Culture* (London: Oxford University Press, 1926):

. . . the Israelite does not distinguish between a living and a lifeless nature. All is living which has its peculiarity and so also its faculties. A stone is not merely a lump of material substance. It is, like all living things, an organism with peculiar forces of a certain mysterious capacity, only known to him who is familiar with it. Thus, like all other beings of the earth, the stone has the quality of a soul, and so also can be made familiar with other physical forces and filled with soul-substance. The earth is a living thing. It has its nature, with which man must make himself familiar when he wants to use it; he must respect its soul as it is, and not do violence to it while appropriating it. (I–II: 155). . . .

. . . earth itself is alive. We know that the Israelites do not acknowledge the distinction between the psychic and the corporeal. Earth and stones are alive, imbued with a soul, therefore able to receive mental subject-matter and bear the impress of it. The relation between the earth and its owner . . . is a covenant-relation, a psychic community, and the owner does not solely prevail in the relation. The earth has its nature, which makes itself felt, and demands respect. The important thing is to deal with it accordingly and not to ill-treat it . . . to deal kindly with the earth, to uphold its blessing and then take what it yields on its own accord. (I–II: 479)

In light of Pedersen's important assertion, I think it fair to say that, biblically speaking, there is an important intermediate point between "pagan animism" on the one hand and "indifference to the feelings of natural objects" or even the total denial of any such feelings on the other. This covenantal midpoint surely offers an important contribution to a planet-respecting attitudinal basis for our relation to our surroundings.

There is further biblical evidence to support this view of nature/ environment. Psalm 148 is widely known and widely read in services both Jewish and Christian. Within Jewish tradition it is a part of every traditional morning service, weekdays, Sabbaths, and festivals. It begins: "Praise the Lord! Praise the Lord from the heavens; praise the Lord in the heights!" Sun, moon, and stars of light are summoned to praise the Divine along with angels and hosts of heaven. It continues: "Praise the Lord from the earth" and summons to praise God sea monsters and all deeps, fire and hail, snow and frost, vapor and stormy wind, mountains and hills, fruit trees and cedars, beasts and cattle, creeping things and winged fowl, along with women and men both young and old and persons from all stations of life. All created things are here adjudged to have at least that minimal measure of sentience that permits joining in this universal hymn to the Creator.

Nor does this tradition end with the Bible. The "Song of the Three Jews," found in the Apocrypha as an addition to the Book of Daniel, is probably a Hebrew composition from the second or first century B.C.E. It too summons all the works of the Lord to sing God's praise and exaltation:

> Bless the Lord, all you works of the Lord,
> sing praise to God and exalt God forever.
> Bless the Lord, you heavens . . .
> Bless the Lord . . . sun and moon . . . stars of heaven . . .
>   rain and dew . . . winds that blow . . . fire and heat . . .
>   dews and snow . . . nights and days . . . ice and cold . . .
>   mountains and hills . . . all that grows in the ground . . .
>   seas and rivers . . . springs . . .
>   whales and all that swim in the waters . . .
>   all birds of the air . . . all beasts and cattle . . .
>   all people on earth . . . all Israel . . . etc. etc.

Let me cite one further example of this continuing strand of Jewish tradition. "Perek Shira, A Chapter of Song," is a fifth- to seventh-century mystical hymn that was first introduced into holiday prayer books by the German Pietists; under the influence of the Kabbalists of Safed it finally became a standard inclusion in traditional daily Hebrew prayer books and can now be found in a fine English version by Barry Holtz in *The Jewish Almanac*, edited by Richard Siegel and Carl Rheins. In it "the entire world of God's creation, the entire cosmos . . . every creature, every living and inanimate thing sings its own special song" in praise of the Creator. Land animals (cows, camels, horses, mules, donkeys, elephants, lions, bears, wolves, foxes, cats, serpents, snails, mice, rats, dogs); winged creatures (roosters, chickens, doves, eagles, cranes, sparrows, swallows, peacocks, storks, ravens, starlings, geese, even vultures); insects (butterflies, locusts, spiders, flies); sea monsters, fish, frogs—all of these and more offer biblical words of praise to their Creator, filling the universe with hymns. Vegetation as well offers songs in celebration of life and creation: the forest tree, the vine, the fig tree, the pomegranate, the palm tree, the apple tree, the stalk of wheat, the stalk of barley, the other grains, the vegetables of the field, and the grasses. Nor does this sense of sentience suffusing the universe end with the vegetative. The heavens, the earth, the desert, the fields, the waters, the seas, the rivers, the springs, day, night, the sun, the moon, the stars, the clouds, the wind, lightning, dew, the rains: they too are members of the universal chorus.

But wait, isn't this the biblical-Judaic tradition whose trust in the predictability of nature and the dependability of the Divine (cf. Whitehead and Einstein) made possible the rise of modern science, predicated as it is upon the uniformity of nature? Indeed, it is the selfsame yet self-varied tradition that propounds both theses. There is a duality of vision, one might say, that affirms both of these poles: creation as sentient and of value in itself and creation as predictable and of value for human purpose and proposal. But just as with our physical sight, where the dual perspective yields depth and richness, so might this dual perspective of the spirit yield a comparably deep and full view of creation. Nor need we despair because it lacks a certain final elegance of unity. The advances of physics in our lifetime, despite the duality of wave and particle hypotheses, surely suggest that created reality may indeed be at depth polar, as the Kabbalists (Jewish mystics) have always insisted. Nor should this be surprising if one keeps in mind Goshen-Gottschein's assertion that multiplicity is the hallmark of creation.

The Good Life

In light of the foregoing, what might we say about the good life? There are four classical Jewish principles that ought at least be summarily stated:

 1. The earth must remain habitable.

> For thus said the Lord,
> The Creator of heaven who alone is God,
> Who formed the earth and made it.
> Who alone established it—
> Who did not create it a waste,
> But formed it for habitation:
> I am the Lord, and there is none else.
>
>   (Isa. 45:18)

 2. We are responsible.

In the hour when the Holy One, blessed be S/He, created the first human being, God took the person and let him/her pass before all the trees of the garden of Eden, and said to the person: "See my works, how fine and excellent they are! Now all that I have created, for you have I created. Think upon this, and do not corrupt and desolate my world; for if you corrupt it, there is no one to set it right after you." (Ecclesiastes Rabbah 8:28, in N. N. Glatzer, *Hammer on the Rock* (New York: Shocken, 1966) (gender revisions by the author)

 3. *Bal tashchit:* do not destroy wantonly. (This rabbinic principle is derived from Deut. 20:19–20 and extends to any wastefulness of earthly resources.)
 4. The will of God is that we should follow the middle way and eat and drink and enjoy ourselves in moderation (Maimonides, Eight Chapters, as summarized by Husik).

Each of these summary statements invites elaboration. Even without further explanation, however, each probably seems reasonable enough, each probably received a nod of agreement, and yet . . . Isn't something missing beyond the explanation? Mightn't our nodding turn out to be an indication of incipient sleep rather than of impending alertness and action? Where is the motivation? What in these eminently reasonable rules moves us? Reason is an indispensable critic and a valuable ally in our seeking after the good, but it is still only an ally; it cannot by itself be burdened with the entire task of stirring us to act.

 We seem to be cut off, detached from nature. As the editors of this book put it, "We have lost our sense of being rooted in a deeper and more encompassing natural order of reality. . . . Religious consciousness and perspective . . . may be indispensable in ameliorating our present situation by helping us to integrate ourselves in a wider (and surely wiser) natural reality . . . a love of the earth that human beings once felt strongly,

but that has been thinned and demeaned." Decades ago, D. H. Lawrence
put it rather strikingly:

Oh, what a catastrophe for man when he cut himself off from the rhythm of the year, from
his union with the sun and the earth. Oh, what a catastrophe, what a maiming of love when
it was a personal, merely personal feeling, taken away from the rising and setting of the sun,
and cut off from the magic connection of the solstice and the equinox! That is what is the
matter with us. We are bleeding at the roots, because we are cut off from the earth and the
sun and stars, and love is a grinning mockery, because, poor blossom, we plucked it from its
stem on the tree of Life, and expected it to keep on blooming in our civilized vase on the
table.

Indeed, we are cut off from those "rhythms of the year" characterized with
such freshness and wonder in another covenant passage from Genesis:

עד כל-ימי הארץ
זרע וקציר וקר וחם
וקיץ וחרף ויום ולילה
לא ישבתו.

"As long as the earth endures, seedtime and harvest, and cold and heat,
and summer and winter, and day and night, shall not cease" (Gen. 8:22).

How recapture those rhythms? How reintegrate our lives with them?
How renew our appreciation for the divine (or cosmic, if you prefer) gift of
nature, the environment? The price for failing to do so is high: "As civilization advances, the sense of wonder declines. Such decline is an alarming
symptom of our state of mind. Mankind will not perish for want of
information; but only for want of appreciation" (Abraham Joshua He-
schel).

This is the core of our task, it seems to me, to help rekindle in all of us
an appreciation of nature, a love for all creation, without which neither it
nor we are likely to survive. So in the remainder of this chapter I want to
share some very specific tangible and visible ceremonies and symbols that
might contribute to our regaining this basic delight in things simple and
wondrous that so abundantly surround us—and so, on to sun and seasons,
soil and spirit!

Seedtime

As I write this, it is autumn in New England, the end of the harvest season;
yet this very autumn is itself the seedtime for certain crops. And so at our
temple we have, over the past two decades, developed the following practices as part of our Sukkot (Feast of Booths) observance. (Cf. Lev. 23:33–35,
41–43.)

At the same time that we celebrate the harvest, we plant some winter rye or winter wheat in the following manner. On the intermediate Sunday of Sukkot, children in our Religious School come outdoors to wave in ceremonial directional fashion the palm frond, willow, myrtle, and citron (cf. Lev. 23:39–40); they eat grapes in the Sukkah (the autumnal harvest booth); they then come to the small patch of prepared soil, each to broadcast a pinch of the wheat or rhy berries, thus seeding a further harvest. The berries germinate and begin to grow while warmth remains in the ground; and against the south facing wall that absorbs the heat of the sun their surroundings are quite benign.

Unpreached yet implicit is a simple teaching: no planting, no harvest; without some human tending, the earth will not provide the bounty on which we depend.

At the same time, there is a focusing of attention on the mystery of the seed. One need not be a devotee of Eleusis to recognize that wonder. In fact, were it not just a bit too cute, I would say that the seed is a seminal image for biblical tradition, where "offspring" is literally "seed" (cf. Gen. 1:11–13; 15:13,18, etc.). As the children (and adults) follow the life cycle of this grain seed through winter dormancy to spring revivification and see the spikelets form and the grain fill out, the life force inherent in a grain of seed is vividly manifest. And as each week we cut some of this growing and ripening grain for use at our Sabbath services between Passover and Shavuot (the Feast of Weeks), our observance of the Counting of the Omer is not only calculational but vegetational (cf. Lev. 23:9–11, 15–16). Thus do seedtime and harvest become harvest and seedtime as we live out religiously the great round of vegetation.

Harvest

The onset of cold weather seems to end the harvest season in New England, and for most of us this is indeed the case. Yet there are some who, in respectful but knowing cooperation with nature, have managed to extend the harvest in an environmentally responsible way. Eliot Coleman's *Four-Season Harvest* (White River Junction, Vt.: Chelsea Green, 1992) is a splendid report and how-to-do-it manual, so this year, to celebrate Sukkot we invited Eliot to come to our temple for the weekend of Sukkot. Friday evening, at the Sabbath service, he and his wife, Barbara Damrosch, spoke on extending the harvest, with photographs of their sun-heated cold frame in central Maine housing thriving lettuce, spinach, and mâche, surrounded by heavy snow drifts. Sunday morning they led a workshop and directed the construction of the portable cold frame now in its first year of use at our temple. We shall not know its full results until later in the winter, but I can well imagine that some of the fresh, leafy produce will find its way to the ceremonial table of Tu Bishevat, the New Year of the Tree, celebrated at full moon of January/February each year.

These particular illustrations of seedtime and harvest are specific to the Jewish religious tradition and our customs at Temple Emanuel, but I am confident that each of our religious traditions can find natural and appropriate way to integrate seedtime and harvest into the rhythms of its liturgical year.

Day and Night

Throughout autumn, the almanac tells us, daylight steadily diminishes by a hardly perceptible couple of minutes each day. On the last Saturday night of October, however, clocks "fall back," and with a startle we find darkness descending distressingly early the following afternoon. It is on the last Friday evening in October, at the beginning of this weekend when we recognize so definitively autumn's approaching end, that our Sabbath evening service for many years has been a Jacob's Lantern Service. Jacob's Lantern is obviously a somewhat whimsical Hebraization of the traditional American jack-o-lantern, and all are invited to bring a carved pumpkin that can shed light throughout the traditional Sabbath service. The service is augmented by appropriate poetry of James Whitcomb Riley, Carl Sandburg, Robert Frost, and others; supplemental music by the organist adds to the hushed wonder, the slight anxiety among the young children, and the subdued joy of all of us as the electric lights are turned off and we spend some quiet minutes amid the golden glow of these illuminated miniature suns, each one a distinct personality also. American folk tradition joins biblical Judaism in a jolly Sukkot Sheni, a second Sukkot, distinctively of this region and this season yet with unmistakable roots in ancient biblical soil as well.

The response of the people? In our small congregation of fewer than one hundred families, more than fifty Jacob lanterns shed light at our service this year, a remarkable turnout indeed! Is it not testimony to the natural vitality of biblical tradition, the wholesome appeal of an American countryside tradition, and the strong desire for integration of our varied traditions with an intimate relation to the rhythms of nature?

Liturgical Garb

In most of our traditions we are familiar with special items of clothing worn during worship. In some cases only the officiant wears, for example, a stole; in Judaism, on the other hand, the *talit* (prayer shawl) and *kippah* (skullcap), if used, may be donned by all worshippers. When conducting communal worship and at other times also, I wear the *talit*, the garment with the fringes on its corners (cf. Num 15:37–40). Most of the time I wear one particular prayer shawl, but for certain special occasions I feel moved to wear a different one, and our Jacob Lantern Service is such an occasion.

This special prayer shawl is one my wife and I happened across in San Cristobal de las Casas, Chiapas, Mexico, when it was simply a rebozo. The attachment of *tzitzit* (fringes), specially tied to serve as a numerical reminder of the Divine, has made this exuberant outpouring of color and fecundity the perfect prayer shawl for such a service. It is also ideal for our special celebrations of the turnings of the seasons as well as certain other occasions.

Why not incorporate in vesture for prayer something of the spirit of Psalm 65:

> You crown the year with Your bounty;
> Your wagon tracks overflow with richness.
> The pastures of the wilderness overflow,
> the hills gird themselves with joy.
> The meadows clothe themselves with flocks,
> the valleys deck themselves with grain,
> they shout and sing together for joy. (11–13)

If hills, meadows, and valleys gird, clothe, and deck themselves to sing together for joy, why not we as well? The spirit of nature so manifest in some psalms invites more than merely verbal inclusion in our services.

The Turning of the Seasons

The succession of the seasons is slow and subtle, and only by astronomical calculation have we been able to designate particular days as marking the turning of the seasons. Yet having determined such dates, we want to mark their appearance in some special liturgical way. Many of our services include special readings for each of the four seasons, yet more is possible. For with our globe spinning, constellations shifting, the sun declining, and the leaves changing, with everything in our universe moving, why should we alone sit motionless amid such cosmic turning?

Back in April 1981, there was celebrated a Hebrew ceremony, held once every twenty-eight years, called the Blessing of the Sun. Details of this mythic occasion must be reserved for another discussion, but I felt the need for some wheel-like object to mark this event. Ezekiel's vision of the chariot (cf. Ezek. 1:15–21; 10: 1 ff) and Arjuria's Krishna-chauffeured chariot (cf. Bhaegavad Gita, chap. 1) may have contributed to this prompting, for religious journeys, like others, involve movement. Whatever the case, I was fortunate to find a religiously sensitive and gifted artist, Karen Frostig, to help design and execute a vibrant sun wheel. Exigencies of space preclude a full explanation of the symbols, some of which are probably self-evident; but the Hebrew words are an acrostic, ascribed to the angel Michael, celebrating the Blessed Creator God, Great in Knowledge, who fixed the sun, sent forth its rays, and established other luminaries in the sky. Fitting, indeed, for a celebration of sun and light.

That occasion was over, and now came the question of what to do with the sun wheel for the intervening twenty-eight years? Put it in a closet where it would be lost to sight and gather dust? That seemed a pity. Besides, who has spare closet space of this size? (The sun wheel is four feet in diameter.) It was at this point that the idea presented itself: why not turn it at each turning of the seasons? So for the past thirteen years, at the Friday evening service nearest to each equinox or solstice, a ritual spinning of the wheel, preceded and followed by appropriate seasonal poetry and accompanied by fitting music, has helped us recapture the rhythms of this vibrant and turning universe in which we are blessed to reside.

Sun

On reflection we find that each one of the previous examples is sun-linked, determined and sustained by the rhythmic energy of that luminary. Should not an environmentally responsible and religiously appreciative life make every effort to utilize more fully, in noninjurious ways, the abundant energy flowing from that source?

A season-bridging constant reminder of this is possible. On the roof of our temple are two panels of photovoltaic cells, silicon derivatives with the capacity to convert sunlight into electricity. Nearly fifteen years ago, in December 1978, we converted the Eternal Light of our temple, the light burning perpetually above the Ark containing the Torah scroll, to solar power. Lack of space precludes a full discussion of the musings and motivations that prompted this shift, among them reflections on present polluting and diminishing energy sources as inappropriate symbols of divine dependability. Religiously, however, the symbolic power and elegance of plugging the Eternal Light directly into the sun (with nighttime mediation, of course, by the storage batteries) have been perceptible and constant these fifteen years, and the increased awareness of the possibilities of solar energy has been environmentally salutory as well. Once again, this particular object of Jewish tradition may invite your reconsideration of a sacred lamp or light within your own tradition as a possibility for this ultimate linkage with the sun as a source of energy and a symbol of nondiminishing divine dependability.

These are among the ways that we found to join the cosmic chorus and recapture the seasonal rhythms. Each autumn we practice these rites, and there are, of course, comparable ways that we celebrate each of the succeeding seasons.

Yet there is one problem that we ought to acknowledge. Even as we move rhythmically with the seasons, with what words do we join the chorus? The traditional words of our synagogue liturgy praise the Creator, the Giver of this universe; yet for many, praise is possible only for the Givenness, not the Giver of our world. Such is the inclusiveness of ritual, however, that the participation in these ceremonies is not precluded by particular verbal formulas. As song and sight, sound and movement, mobilize our energies and direct our attention to the constantly renewed wonders of our existence, the celebrative sense of gratitude seems not merely to defy but to dispense with words literally understood: the saying, the swaying, and the singing transcend the words that initiated our actions.

Thus do we celebrate, and come to cherish all the more, the fresh gifts of each season. And from the cherishing must surely come our ever deepening caring for this planet.

Be this God's will.   כֵּן יְהִי רָצוֹן.
        Amen.   אָמֵן.

A Contemporary Evangelical Perspective

Some Kinds of Knowledge

Knowledge of How the World Works

Condors, Children and Other Parts. Professor Timothy Weiskel has related how he and Professor Harvey Cox, also of Harvard, engaged in a debate centered around an ethical question. The question was this: "If during the 1993 Chaparral fires in southern California you came upon a situation where you could save a California condor or a small child and the time available limited you to rescuing one, which one would it be?" They then agreed to make the question more difficult: "If during the recent Chaparral fires in southern California you came upon a situation where you could save the last remaining individual of a species of weed or a small child and the time available limited you to rescuing one, which one would it be?" Clearly, these questions raise the ethnical question "What is right?" but I would like to use these questions to get to equally as important a question, "How does the world work?"

Knowledge of the Chaparral ecosystem would show that it is one whose continued existence is assured by fire. The very nature of this ecosystem is that it is maintained and purified of invading Eurasian plants by fire; fire is essential for its persistence, and so too is fire necessary for the persistence of the "weed" in question. When armed with such knowledge, the answer to the question clearly is "Save the child, for in saving the child you also save the plant (which survives by fire). If one, believing that saving a species was more important than saving an individual human being, would quickly dig up the plant to rescue it, *both the plant and the child* would be lost. Moreover, if we were to take the first question rather than the second, we also would save the child, for the California condor has wings and can lift from the ground clear of the fire and fly away, thus saving itself.

What is important for us to learn from this is that knowledge of what is right, although clearly important, is not sufficient. We also must have knowledge of how the world works. Saving the weed's life would likely result in losing it, and "losing" the weed's life would likely result in saving it. Action must be guided not only by ethics but also by knowledge of the system.

Systems of Interacting Parts: Fishing and Knowledge of Systems Dynamics. In a workshop conducted by Dennis Meadows, we participated in a fishery simulation game. Each of five teams sought to maximize our catch of imaginary fish in the imaginary deep sea and coastal waters by continuing to add ships to our fishing fleets. As the money accumulated, we all purchased more ships, and then—after several rounds of play—the fishery collapsed. What we discovered, to our chagrin, was that our knowledge of the cause of the collapse—overfishing—came too late to do anything helpful. Our single-purpose mind-sets had driven us to maximize production without looking at the system as a whole. A potentially sustainable fishery was devastated.

Each of our teams had all the understanding recorded in writing that would have allowed us to manage the fishery on a sustained-yield basis, but all of us were so busy making money that none had the time to think through the consequences of our actions. After the great collapse, we found that it would have been impossible for us to take needed corrective action even at the first hint of a decline in the fishery, for by then it was already too late. We discovered, of course, that we not only had to have knowledge of the system but also that we had to have this knowledge organized in such a way that we could predict the degrading consequences of our present practices on the future; we had to put the knowledge together to allow us to see the results of our actions several years ahead.

This brought us to understand that we not only had to know the present state of the system with which we were interacting, we also had to know the system as a dynamic set of interactions between entities, and we had to know how these interactions would play out in the long run. We had to have "systems dynamics knowledge" that would allow us to understand interactions and their consequences now and into the future. If we were to apply such systems knowledge of the Chaparral ecosystem in which we imagined our condor and child, we might never have provided a child to be present in the Chaparral in the first place, since our systems knowledge would have allowed us to predict the high probability of a fire before it occurred, which in turn might have prevented us and the child from being present in such a potentially incendiary situation.

I bring these first two points to our attention because, prior to dealing with ethical teachings, it is necessary to emphasize the vital importance of having knowledge of the world in which we apply these ethics. We must know how the world works. But there is more to be said about knowledge than "knowing how the world works."

Knowledge of Method and Technique

If we are to put into service knowledge of how the world works and knowledge of ethics, it also is important to have knowledge of the methods and techniques available to act on this knowledge. This includes methods of leaving things alone and protecting things from disturbance, practices of soil and water conservation, stewardship of biotic communities and ecosystems, techniques of restoration ecology, and techniques for shaping and reshaping things to meet the needs and wants of human life.

Which much of past knowledge along these lines has been knowledge of the technology of exploitation, increasingly we employ, and need to employ, techniques of preservation and restoration. With use and abuse of the land and creatures now so widespread, knowing such preserving and restoring techniques is particularly important. Nowadays, an immense amount of effort and thorough knowledge of legal and other procedures are needed even to keep things as they are.

While we must know how the world works and what methods of techniques are available, still more needs to be said about knowledge. To get to this, I use the case of the hydrologic cycle.

Knowledge of the Sustainer: The Hydrologic Cycle

Water in the world about us is cycled and recycled. Taken up by animals, it is released through breathing, sweating, panting, and ridding of wastes— finding its way to the atmosphere—or through the route of sewage treatment plants back to rivers and streams. Taken up by the roots of plants, some is pumped up through the bundles of tubing in the roots, stems, and leaves of plants and back to the atmosphere, while some is used together with carbon dioxide to make the stuff of life that, after use by plants and animals as building materials and fuels, once again is released to the atmosphere. The water that goes into the atmosphere joins water evaporated from lakes, streams, soil, and other surfaces, eventually forming rain, sleet, or snow that again waters the face of the earth. Some runs off to streams and other surface waters, again to evaporate and re-form as the clouds from whence it came. Some percolates through the soil back to roots of plants, and some slips past these roots to enter the groundwater, to be pumped by wells for human use or to emerge and eventually to be returned to the clouds again. As water is evaporated or transpired to the air, almost everything it contained is left behind—a sweet distillation expressing a bountiful love of God for the world. And the clouds—great condensations of distilled watery vapors—rain it all down again to water the earth. The hydrologic cycle.

Of course, we can find this cycling of water in the biosphere described in textbooks, but it also is described in hymns and psalms. As the hymn has it:

> Thy bountiful care what tongue can recite?
>   It breathes in the air, it shines in the light;
> It streams from the hills, it descends to the plain,
>   And sweetly distills in the dew and the rain.
> O worship the King, all glorious above,
>   O gratefully sing His power and His love.
> Your ransomed Creation, with glory ablaze,
>   in true adoration shall sing to your praise!

And as Psalm 104:10–13 puts it:

> He makes springs pour water into the ravines;
>   it flows between the mountains.
> They give water to all the beasts of the field;
>   the wild donkeys quench their thirst.
> The birds of the air nest by the waters;
>   they sing among the branches.
> He waters the mountains from his upper chambers;
>   the earth is satisfied by the fruit of his work.[1]

Cycles upon cycles . . . cycles within cycles . . . cycles of cycles—the Creation is permeated with cycles, and each of these is empowered by energy poured out from the sun. The workings of ecosystems rely upon all of this cycling. The biosphere—that great big envelope of life that covers the face of the earth—is composed of prairies, oceans, forests, lakes, glades, woodlands, brooks, and marshes; it is composed of ecosystems. Waubesa Marsh, the big wetland on which I live, is one of these ecosystems. It, like every other ecosystem on earth, has its plants, animals, soils, and climate: there are the sandhill cranes, whose six-foot wing spans, seventy-year life spans, and bugling calls seemingly command the great marsh; there are the iron bacteria, whose tiny size and very short lives would escape our notice except for the oil-like film they create over the quiet waters; there is the deep peat soil that at the lake edge extends to a dizzying depth of ninety-five feet and holds within itself a record of pollens, seeds, and other remains that define its long history; and there is the ebb and flow of water: its coming in from flowing springs and falling rain, its leaving by flowing streams, by transpiration through the pores of wetland plants, and evaporation from surfaces of land and water. These creatures and their interactions, and much more, make up the wetland ecosystem.

Although it might not first meet the eye, ecosystems are places of immense ecological harmony. Not every creature plays the same tune, so to speak, but in so many ways they all are in tune with each other—in harmony, in polyharmony. A great marsh, at first seemingly unstructured and disordered, is in time discovered to be a highly ordered system in which each creature interacts with the other creatures to form an integrated whole. And what is true in regard to wetlands is true of forests and prairies, lakes and deserts. Each is a kind of symphony, and the biosphere is a symphony of symphonies, where all creatures great and small are so related with each other that they continue to produce after their kinds generation after generation, continue to maintain and sustain the living fabric of the biosphere, continue to bring forth life from death, continue to cycle and recycle the basic stuff of Creation—all powered by our star, the sun. The various ecosystems of the biosphere are provided with everything needed for their continuance through the years and generations, everything needed for their creatures to interactively sustain the whole system in which they have a part. Again, while all of this can be, and is described scientifically, it also is described in song. Stuart Hine, in 1953, put it this way:

> O Lord my God! When I in awesome wonder
> Consider all the works Thy hand hath made . . .
>
> When through the woods and forest glades I wander
> and hear the birds sing sweetly in the trees,
> When I look down from lofty mountain grandeur
> And hear the brook and feel the gentle breeze.
>
> Then sings my soul, my Savior God, to thee:
> HOW GREAT THOU ART! HOW GREAT THOU ART!

Before I come to describing the different kind of knowledge toward which I am leading, let us continue with water. In the cycling of water on earth we know that some water percolates through the soil to the groundwater below, and that water is what eventually supplies the flowing springs that feed the wetlands, lakes, and ravines; we call this *percolation*. We also know that some water is returned to the air by evaporation from the surfaces of water, land, and organisms and from transpiration through the pores of leaves; we call this *evapotranspiration*, or simply ET.

We already have noted that as water is evaporated or transpired to the air, almost everything it contained is left behind—"a sweet distillation," we called it. Evapotranspiration is one important provision for purifying water in the world.

Another important provision for purifying water is percolation. We already noted that percolation is the movement of water downward through soil and that such movement eventually brings it to the groundwater. In many water treatment plants in our cities, water is treated by having it percolate through beds of sand; this results in removal of many impurities in the water. In similar fashion water that percolates through the soil is treated but usually over much greater distances through soil and rock. The result is that by the time we pull up the groundwater to our homes by means of our wells or the groundwater emerges as springs, it usually is fit to drink. Percolation and the movement of groundwater through aquifers of soil and rock are important provisions for purifying water in Creation.

Still other important provisions are the brooks, streams, and rivers. At normal levels of waste input in natural ecosystems, these flowing waters and their living inhabitants remove the impurities so that by the time water moves a few miles downstream the impurities put in upstream are largely removed. So the processing of water by flowing streams is another important provision for purifying water in Creation.

ET, percolation, and flowing rivers—and there is yet one more: wetlands. The great marsh where I reside as I write this, and other wetlands of many types across the globe, serve as water purifiers under natural conditions. Thus, when water that has picked up eroded soil as it flows across upland areas enters wetlands, the soil particles are filtered out. And in many instances dissolved chemicals also are taken up by wetland plants. The result is that water entering rivers and lakes by way of wetlands are cleaned up before entering, and that is important for keeping flowing waters and lakes habitable for other life.

There is wonder in all of this! All of us know what water is. And yet it is so common in most of our lives that we take it for granted. So we need to be reminded that it is what often is called "the universal solvent," meaning that it dissolves practically anything. And this fact should cause us to think about how water can ever be purified. Since it is the universal solvent, should it not always be contaminated with dissolved materials from everything through which it passes? Water also is the only major liquid substance in the world and as such flows from place to place, bringing with it all sorts of particles held in suspension; so should it not be contaminated with all sorts of suspended material. Should not the world be a "big soup"? The answer, we have found, is no because of the natural "distillers," "filters," and "extractors" of Creation. There is remarkable provision in Creation for the production of pure water; once having been contaminated by sediments and dissolved substances, it is made pure again and again and again! This provision makes a vitally important contribution to the fruitfulness and abounding life of earth, of which the psalmist sings:

> Your Spirit O Lord, makes life to abound.
> The earth is renewed, and fruitful the ground . . .
>
> God causes the springs of water to flow
> in streams from the hills to valleys below.
> The Lord gives the streams for all living things there,
> while birds with their singing enrapture the air.
>
> Down mountains and hills your showers are sent.
> With fruit of your work the earth is content.
>   (Psalm 104:10–13)

An early creed of 1561 describes the different kind of knowledge toward which I am leading, with these illustrations. Here is the statement of the Belgic Confession, based on Romans 1:20, Psalm 19.1 4, and Acts 14:17:

> Article II: *By What Means God Is Made Known to Us*
>
> We know him by two means:
> First, by the creation, preservation, and government of
> the universe;
> which is before our eyes
> as a most elegant book,
>> wherein all creatures,
>> great and small,
>> are as so many characters
>> leading us to see clearly
>> the invisible things of God,
>>> even his everlasting power
>>> and divinity,
>>> as the apostle Paul says (Romans 1:20).
>
> All which things are sufficient to convince men
> and leave them without excuse.
>
> Second, He makes Himself
>> more clearly and fully known to us

by his Holy and divine Word,
that is to say, as far as is necessary for us
to know in this life,
to His glory
and our salvation.[2]

The principle biblical teachings upon which this is based are Psalm 19:1–4:

The heavens declare the glory
of God.
the skies proclaim the work
of his hands.
Day after day they pour forth
speech;
night after night they display
knowledge.
There is no speech or language
where their voice is not
heard.
Their voice goes out into all
the earth,
their words to the end of the world.

Romans 1:20: "For since the creation of the world God's invisible qualities [God's eternal power and divine nature] have been clearly seen, being understood from what has been made, . . . so that people are without excuse."

The third kind of knowledge is the knowledge that the world conveys about its Creator. The world breaks forth with a marvelous testimony, one that is so powerful that it leaves everyone without excuse about knowing of God's everlasting power and the fact that God is God, that God is divine. I remember in my youth savoring the words of Article II of the Belgic Confession because it affirmed in a deep theological way the worth of my continuous observation of animals and plants in the city, the city dump, and in the country beyond. Today, as I write this, on Waubesa Marsh, the heavens continue to tell the glory of God, and the creatures continue to pour forth their testimony to God's eternal power and divine majesty. It is a drizzly day in early spring; the marsh seems expectant of the great burst of life that is upon us; the geese call above, and six sandhill cranes with their clangoring calls announce the arrival and revival of life on the great marsh. Most of us have had awesome experiences in this world. Perhaps we have stood at the edge of a great canyon or at the feet of giant trees in an ancient forest or in the center of a great storm. Perhaps we ambled on a flowering meadow in the quietly lifting mists of the morning dew. Perhaps this elicited in us the hymn, "How Great Thou Art." The Scriptures say that this elicitation, or something like it, will happen to all of us.

Unprecedented Knowledge; Unprecedented Degradation

Our published scientific knowledge of the world and of its conservation and stewardship are unprecedented. Yet never has there been greater degradation and destruction of life and environment. The present environmental state of the world constitutes the most serious threat to the biosphere since the origin of life on earth. The biosphere is being seriously degraded by human activity, summarized in what can be described as seven degradations of Creation: (1) *alteration of the earth's energy exchange with the sun* that results in *global warming* and *destruction of the earth's protective ozone shield*; (2) *land degradation* that reduces available land for creatures and crops and destroys land by erosion, salinization, and desertification; (3) *water quality degradation* that defiles groundwater, lakes, rivers, and oceans; (4)*deforestation* that each year removes primary forest the size of Indiana and degrades an equal amount by overuse; (5) *species extinction* that finds more than three species of plants and animals eliminated from the earth *each day*; (6) *waste generation and global toxification* that results in DDT in Antarctic penguins and pesticides in a remote lake on Isle Royale in Lake Superior; and (7) *human and cultural degradation* that threatens and eliminates long-standing human communities living sustainably and cooperatively with Creation, together with the loss of long-standing garden varieties of food plants.[3]

This diminishing of the integrity of the biosphere—these seven degradations—reflect a crisis in the whole life system of the modern industrial world including nature and the human culture it supports and sustains. Upon probing the causes, we find them to rest with ourselves and the way we are living. We are the cause of the degradations that affect the environment and ourselves.

The words of the prophet Hosea to the people of his time are sobering to ours today:

> There is no faithfulness, no love,
> no acknowledgement of God in the land.
> There is only cursing, lying and murder,
> stealing and adultery;
> they break all bounds,

and bloodshed follows bloodshed.
Because of this the land mourns,
and all who live in it waste away;
the beasts of the field and the birds of the air
and the fish of the sea are dying. (Hos. 4:1–3).

Our first responses to environmental degradations in the 1970s were legal and technical, but we have found them wanting. Although perhaps necessary, they are not sufficient. The earth's ecological deterioration is at heart a matter of human attitudes toward the earth and life in general. An ethical response is vital. And the needed ethical response must be rich and full, touching every level of our being, one that addresses what we consider to be of ultimate importance in our lives and how we think we ought to live, one that reflects morally on how we understand and relate to nature.

Ethics the Missing Element

Recognition of ethics as the missing element has been indicated by environmental scientists who move beyond description to condemn Creation's destruction, by philosophers who explore environmental ethics, by inventors of new religions, by writers of ecosystem ethics, and by engineers who inject environmental ethics into curricula.[4] This recognition is also signified by explorations in the world religions for environmental teachings, such as the Assisi Declarations that proclaim "destruction of the environment and the life depending upon it is a result of ignorance, greed and disregard for the richness of all living things" (Buddhist); that we "repudiate all ill-considered exploitation of nature which threatens to destroy it" (Christian); that we should "declare our determination to halt the present slide towards destruction, to rediscover the ancient tradition of reverence for all life" (Hindu); that "now, when the whole world is in peril, when the environment is in danger of being poisoned and various species, both plant and animal, are becoming extinct, it is our . . . responsibility to put the defence of nature at the very centre of our concern" (Jewish); and that people as God's trustees "are responsible for maintaining the unity of His creation, the integrity of the Earth, its flora and fauna, its wildlife and natural environment" (Muslim).[5]

All these indicators point to a searching for a way to live rightly in the context of growing environmental degradation, to work and live respectfully and restoratively in Creation. The good life, we are finding, is not merely a matter of the person; it comes not from alienation from the biosphere or from biospheric degradation. The good life has its roots in the goodness of Creation. Degradation of Creation erodes the good life. Thus, the instillation into human hearts the resolve to live rightly on earth goes beyond the person and the personal to embrace the biosphere. The good life depends for its support upon the good earth. Any pursuit of the good life that degrades the earth finds its own destruction.

Not a Mere Ethics but an Appropriate Ethics

As our interactions with the ecosystems of which we are part teach us the consequences of their degradation, as we realize the limitations of law and technique, we find we need not *mere* ethics but—since ethics operate among thieves as well as saints—ethics directed at preserving, maintaining, and restoring the integrity of Creation.

This, of course, leads us to ask, "By what standard our ethics?" For this chapter I have taken my lead in part rom Hugo Grotius, the founder of international law, who was driven by the context of his times to develop an ethics for international relations in his *Law of War and Peace* in 1625.[6] This work, "quite as much a treatise on religion and ethics as on law,"[7] is one that takes the Bible seriously. In following his lead, I will do so also, for the following reasons: (1) it provides a long-standing written and canonical ethical system that has served human societies for more than three thousand years, with an influence that continues to be substantial and sustained; (2) it has contributed to the ethics of communities and civilizations that have persisted through this period and thus has "survival value"; (3) it has, as Grotius recognized, validity even apart from its biblical underpinnings;[8] and (4) it recognizes that belief in God continues strongly into the present In a large part of society.[9]

Evangelical Belief

In the present day there are many religious people who take the Bible seriously as ruling life and practice and believe that the teachings of the Bible should be publicized. Although the term has been defined in various ways, "evangelical" is perhaps the adjective to apply to these people. At its base, *evangelical* references two basic tenets: (1) the Bible is a book that is to be taken very seriously in guiding what is believed and is practiced, and (2) its message should not be selfishly kept. Usually, such evangelicals are called Christians; thus the term "evangelical Christians." Some who do not call themselves evangelical are so by this definition, and some who call themselves thus are not; but by and large, those churches and denominations that identify themselves as evangelical are so and many who do not so identify themselves also are. The etymology of the word supports the definition used here. Derived from the words *eu* (Gr.: true) and *angelis* (a messenger, or bearer of news), *evangelical* refers to those who bring good news.[10] The good news they bear is the testimony of the bible on how rightly to live on earth.

When a religious tradition professes that the Bible is a book that is to be taken very seriously in guiding what is believed and practiced and that its message should not be selfishly kept, there are direct and vast implications for its conception of the nature of God, and of what constitutes the good life, and its view and attitude toward the environment. In our ecological times this implies major transformation in the way believers in this tradition live their lives and relate to the environment. People of this tradition measure themselves against the standards of the Scriptures. As they deviate from this standard for faith and practice, adjustment and conversion are necessary.

The evangelical Christian tradition, along with other traditions of the Book, recognizes that human beings, even while piously professing beliefs, may go astray as individuals, families, communities, and nations. It recognizes the tendency of people to become alienated from God, from neighbor, and from Creation. Thus, it tries to remain open to God's call to get back on course. And of course, this tradition includes not only the Hebrew Bible of the Judaic faith but also the New Testament. Because of the fact that the majority of the biblical teachings on Creation and its stewardship come from the Hebrew Bible, there are many points of similarity between Judaism and Christianity in environmental teaching. And since portions of the Bible are authoritative for Islam, there are similarities with Islam as well. And because of the largely monotheistic religious traditions of many Native American peoples, including the belief in the Creator, there are similarities with this oral faith tradition as well.

Believing that God is the creator, owner, and sustainer of the whole of Creation, evangelicals (along with many others) come to see environmental degradation as a spiritual crisis. It is a spiritual crisis because God is of ultimate importance, and all creatures are created and sustained by their Creator. How one relates to Creation reflects how one respects and honors the Creator. Honoring the Master Artist while trampling the Master Artist's works is an intolerable hypocrisy that must be corrected. Similarly, seduction by immediate pleasures and goods of the world alienates people from God and Creation. In the words of Vaclav Havel, "A person who has been seduced by the consumer value system, whose identity is dissolved in an amalgam of the accoutrements of mass civilization, and who has no roots in the order of being, no sense of responsibility for anything higher than his or her own personal survival, is a demoralized person." A society so seduced is a demoralized society.

Emphasizing the importance of the Bible, however, does not mean that the Bible is the only source of knowledge, for the Bible itself commends the Creation to people as another source. In this and some other traditions that derive from the Book, there is a "two books theology" that professes two ways by which we know God: one through the things and the relationships God has made in Creation and one through what is written in the Scriptures. And these two teachers—God's Word and God's world—engage in dynamic, interactive teaching. Thus, in this tradition one does not build on a flood plain of a river expecting that god will prevent flooding. Neither does one consume carcinogens and expect to be immune from cancer. Thus, human beings must not only steep themselves in God's Creation. If they fail to be students of both the Word and the world, people will become disoriented physically, spiritually, and morally, and their bringing of praise to God will correspondingly diminish.

Three Key Biblical Principles for Honoring God, Caring for the Environment, and Living the Good Life

With specific regard for the environment and the Creation, the nature of the good life has at its core three biblical principles: earth keeping, sabbath, and fruitfulness. Interactions between these three components provide the integration that make the whole a kind of symphony. All components, each maintaining its integrity, interact integratively and harmoniously to form the whole of what is real.

Earthkeeping Principle: We must keep Creation. As God keeps believing people, so should God's people keep Creation. The rich and full keeping invoked with the Aaronic blessing (Num. 6:24) is the kind of rich and full keeping that people should bring to the garden of God—to God's creatures and to all of Creation. Human relationship to Creation must be a loving, caring, keeping relationship. When we *keep* the Creation (Gen. 2:15), we make sure that the creatures under our care and keeping are maintained with all their proper connections—connections with members of the same species, with the many other species with which they interact, with the soil, air, and water on which they depend.

Sabbath Principle: We must give Creation its sabbath rests. As human beings and animals are to be given their times of sabbath rest (Exod. 20, Deut. 5), so also the rest of creation (Exod. 23, Lev. 25–26). People, land, and the creatures of Creation must not be relentlessly pressed. As people observe the Sabbath of the week to help get "off the treadmill" of continuous work, to help get things together again, so too should people observe the sabbath for the land. In keeping with the teaching of Jesus and other Jewish teachers that the Sabbath is made for the ones served by it—not the other way around—this means giving creation necessary rests, intentional nurture, and active restoration. The land and the creatures must be protected from relentless exploitation, must be given what is needed for rejuvenation and getting things together again. The sabbath for the land includes not only agriculture but all of Creation—including our use of water and air, as we discharge into them our exhausts, smoke, sewage, and other things we throw "away." Failure to give the land and the creatures their needed rest eventually will result in people no longer being supported by land and creatures.

Fruitfulness Principle: We must preserve Creation's fruitfulness. People may enjoy the fruit of God's Creation but must not destroy its fruitfulness (Ezek. 34:18). As God's fruitful work brings fruit to Creation, giving to land and life what satisfies, so too should ours. As God provides for the creatures, so should we people who were created to reflect God whose image we bear. Imaging God, we should provide for the creatures, and, with Noah preserve the fruitfulness of earth's creatures, especially those threatened with extinction. Preserving Creation's fruitfulness preserves biotic species whose interactions with each other, and with land and water, form the fabric of the biosphere. Our fruitfulness (Gen. 1:26–28) should not be at the expense of the fruitfulness of other creatures (Gen. 1:22).

The Components: God, the Environment, and the Good Life

This brings us to an evangelical perspective on God, the environment, and the good life. The approach I am using in this chapter is first to identify the three components: God, environment, and the good life; next, to address the interactions between these components; and finally, to set forth the understanding of these components together with their integrative interactions as forming the base for a rich and full response to Creation and the environmental crises that beset it.

The Nature of God

God Is Creator and Owner of All Creation. In the Judaic, Islamic, Native American, and Christian view, "The heavens declare the glory of God, and the firmament showeth God's handiwork" (Ps. 19:1); and "The earth is the Lord's and all it contains, the seas and all that dwell therein" (Ps. 24:1). There generally also is affirmation of the message of Romans 1:20, namely, that the testimony of God in Creation is so apparent as to leave all people in all ages without excuse but to know God's everlasting power and divinity. Relating to these generally shared beliefs is the underlying one, that God is the Creator of all things (Gen. 1, 2). God is thus prior to and other than Creation, yet is personally and intimately involved with us, other creatures, and all Creation. And thus the confession: "You alone are the Lord. You made the heavens, even the highest heavens, and all their starry host, the earth and all that is on it, the seas and all that is in them. You give life to everything and the multitudes of heaven worship you" (Neh. 9:6). God is the Creator of all, God is the owner of all, and what God has created bears convicting witness to God's eternal power and divinity.

God is Righteous and Just. Moreover, the Creator of all things is eternally and consistently just and righteous, with all of god's works deeply rooted in God's law —God's Torah. "Thus," as the fourth century's *Genesis Rabbah* puts it, "the Holy One, blessed be He, consulted Torah when He created the world.[11] While Torah may first be understood as the decalogue (Exod. 20, Deut. 5), it is understood more completely as the first five books of the Bible and even more completely as the law whereby God orders and sustains the whole of Creation. Thus, God's justice, Torah, and Creation are interrelated. Torah is present with God before the creation of the universe and underlies the order, integrity, and goodness of Creation; it is by Torah that God created, is creating, and is sustaining all things. The personified Wisdom of Proverbs 8 is viewed in *Genesis Rabbah* as Torah; she is brought forth and fashioned by God as the first of his works, before the creation of the world; she is the tutor at God's side as God creates the world.

The Lord brought me forth as the first of his works, before his deeds of old; I was appointed from eternity, from the beginning, before the world began. When there were no oceans, I was given birth, when there were no springs abounding with water; before the mountains were settled in place, before the hills, I was given birth, before he made the earth or its fields or any of the dust of the world. I was there when he set the heavens in place, when he marked out the horizon on the face of the deep, when he established the clouds above and fixed securely the fountains of the deep, when he gave the sea its boundary so the waters would not overstep his command, and when he marked out the foundations of the earth. Then I was the craftsman at his side, I was filled with delight day after day, rejoicing always in his presence, rejoicing in his whole world and delighting in mankind. (Prov. 8:22–31)

Reflecting God's justice and righteous, the Creation that God intends is a symphony of all creatures in harmonious relationship among them and with their Creator.

God Is Sustaining and Loving. God pours forth sustaining love for the world. God is richly involved with people, other creatures, and all Creation while wholly other than Creation and not to be confused with it. The Creator cares for people as well as sparrows; the Lord plants the Cedars of Lebanon and renews the face of the earth; the sustainer gives prey to the lions and satisfies the land and its creatures. (Ps. 104).[12] God's love for the world is so great that it even is self-giving (John 3:16).[13]

As creator and owner and sustainer of all, God gives food to the creatures at the proper time and takes away their breath as the creatures become the life-sustaining food of other creatures (Ps. 104:28–29), reserving the Artist/Creator's exclusive right to bring life from death in the trophic symphony of God-ordained food webs and energy transfers.[14]

Following its recounting the Creator's sustaining love, Psalm 104 concludes: "May the glory of the Lord endure forever; may the Lord rejoice in his works" (Ps. 104:31).

God Is Judge and Reconciler. God, according to the Scriptures, gives human beings choice. Creating people not as automata or puppets, God endows human beings with the ability to make up their own minds, and even the ability to go their own way. But the choice with which human beings are endowed comes with the admonition "Choose life!" (Deut. 30:19). God commands people to "love the Lord your God," walk in God's ways and keep God's law (Deut. 30:16) and by doing so they will live safely in the land (Lev. 25:18). However, if they fail to keep God's law, pressing the land relentlessly, they will be driven off the land, and then the land will receive its lawful rest (Lev. 26:14–35), "the rest it did not have when during the sabbaths you lived in it" (Lev. 26:35).

As in human courts of justice, the Creator calls upon witnesses: "This day I call heaven and earth as witnesses against you that I have set before you life and death, blessings and curses. Now choose life" (Deut. 30:19). In the wake of human choosing to go their own way, of human choice to violate God's law, is God's ultimate judgment and God's ultimate refinement of creation and God's reconciliation of all things. "But who can endure the day of his coming? Who can stand when he appears? For he will be like a refiner's fire or a launderer's soap" (Mal. 3:2). "Surely the day is coming; it will burn like a furnace. All the arrogant and every evildoer will be stubble, and that day that is coming will set them on fire" says the Lord Almighty (Mal. 4:1)—a fire that in a later prophecy will result in the works of Creation being discovered anew (2 Pet. 3:10).

Building upon this biblical message is that of Colossians 1:19–20 of "the firstborn over all Creation": "For God was pleased to have all his fullness dwell in him, and through him to reconcile to himself all things." The earth and its creatures, degraded by human actions fueled by arrogance, ignorance, and greed, is to be reconciled—to be made right again—with God's Creation being uncovered and disclosed by the removal of the degraders and their degradation of Creation.

God does not let bad choices by people destroy Creation, but in the exercise of law and justice, God purifies Creation, making it like new. God is judge and reconciler: the destroyers of earth are destroyed (cf. Gen. 6:11–13, Rev. 11:18), and God's wonderful works are uncovered fresh and new to the end that God's justice prevails and that all creatures and the whole Creation return to God their praise.

Implications for Environment and the Good Life. The implications of the character of God in relation to people and the rest of Creation are direct and vast for what constitutes the good life and attitudes and behavior toward nature and the environment. In the pursuit of the good life, this means that human beings should be obedient to the law by which God ordains the whole Creation and all its interacting creatures and that they should be imaging, honoring, and glorifying God in their care and keeping of each other, of other creatures, and the whole of Creation. In the pursuit of the good life this means that in their attitude and behavior toward Creation, human beings should keep it, give it its rests, be creative in their life and work, enjoy its fruits, and preserve its fruitfulness.

The Nature of the Environment

"Creation" and "Environment." There is no word in the Hebrew Bible for "the environment." Surprising for most people today is the fact that neither does the Hebrew Bible have a word for the Creation. And the New Testament refers to the Creation very sparingly. In the biblical view, there is nothing "apart" from us human beings that we can call "the environment" or even "the Creation." The people of the Book view themselves as creatures created by God and thus are part of Creation, not apart from Creation. Human beings are creatures too.

Simply put, there is God and Creation, and people are part and parcel of Creation. This is not to say that people do not have a special task; in the biblical view they do. "Look at behemoth, which I made along with you," declares the Creator to Job (Job 40:15). Thus, to the extent that "environmentalists" separate themselves from the environment, seeing it as separate from themselves—something you can "save" independent of themselves—the term "Christian environmentalist" is an oxymoron. In the biblical view there is Creation in all is wholeness, which when one is fully incorporated into makes it unnecessary even to have a word for it. Creation is all there is, besides the Creator, and thus it really never need be referred to as object. Thus, the nature of the environment in the biblical view is that it is a component of the Creation. Furthermore, care for Creation does not distinguish or prioritize between caring for people and caring for the environment. Caring for Creation always means caring for the whole system, of which spotted owls, furbished louseworts, snail darters, and human beings are interacting components. Thus, Creation consists of "all things," rendered in the Greek New Testament as *ta panta* (cf. Colossians 1:20).

Creation Is Good. In the Creation account recorded in Genesis 1, the created works are repeatedly declared to be "good." While there has been debate about the impact of the Fall (Gen. 3) on Creation's goodness, it is clearly the teaching of the scriptures that this goodness remains sufficient to allow the Psalmist to declare that the heavens declare the glory of God (Ps. 19:1) and for the Apostle Paul to state that Creation's testimony clearly convicts people of God's divinity and eternal power (Rom. 1:20). The goodness of Creation, however, must be distinguished with what we might call "goodiness." God's goodness can even be seen in the power of the storm as it comes up over the Mediterranean Sea and rips into the forests of Lebanon (Ps. 29) and in God's provision of prey for the lions (Ps. 104:21). God's declaring Creation to be good apparently means the kind of symphonic harmony we see present in Creation, including the trophic dynamic ecology of food webs and the remarkable strength and adaptive complexity of the massive hippopotamus (Job 40:15–24). This goodness of Creation exhibits the goodness of the Creator, who is its author and sustainer.

Creation Has Intrinsic Capacity to Heal. Given the opportunity, creatures, ecosystems and Creation have built into them the capacity to heal themselves, within limits. In the case of people and animals, the physicians or veterinarians do not do the healing but set the conditions for healing to take place; the body does the healing. In the case of soil building, the soil scientists or farmers do not build the soil but set the conditions for building the soil. The creatures, ecosystems, and the Creation as a whole has certain God-given abilities of self-regulation, self-restoration, and healing—all for maintaining and restoring Creation's integrity.

Creation as Teacher. In addition to the Bible, Creation also is a great teacher, particularly of the ecological principles whereby it is ordered. In reading Creation as well as the Bible, there is an organic whole—something we expect from their Author, who is characterized by justice, righteousness, and sustaining and reconciling love. As we study both the Word and the world, we inform our worldview; we increasingly operate with "the law written upon our hearts." From the perspective of the New Testament, we increasingly transform our minds to that of the One through whom the whole world is created, sustained, and reconciled. Such transformation of our minds to that of Christ brings people increasingly closer to imaging God in God's care and keeping of people and the rest of Creation, in God's love for the world. Through such transformation we increasingly give hope to Creation and its eager expectation of the coming children of God.

The Material World Is Good. The Bible affirms the importance of the material substance of Creation. Adam is from *adamah* (earth); we are dust and to dust we will return. Moreover, in New Testament teachings, God is "made flesh," takes upon flesh and blood, and dwells among material creatures—becomes *incarnate* (literally, "in the flesh"). The material Creation, brought to a state of gloom in the death of its incarnate Creator, is brought to joyful celebration by Christ's bodily resurrection; Christ's resurrection vindicates Creation.[15] And ultimately, after refining by the refiner's fire—after destroying those who destroy the earth—the material Creation will flourish once again, with real rivers, real lions, and real lambs. The importance of the material is emphasized in the New Testament in part to counter the Gnostic idea of the material being evil and consequently something to be denigrated.

Creation and the Creatures Have Intrinsic Worth. The creatures elicit the praise, pride, and protection of the Creator, irrespective of any utilitarian purpose. Thus the noneconomic animal species are saved from the flood (Gen. 6–9); the hippopotamus and other creatures are held up proudly by the Creator as masterpieces wholly apart from commercial worth or value as pets (Job 40:15–24). Creatures have value simply because they are the works of the Creator.

Creation Is an Administrator of Retributive Justice. Finally, in the biblical view, Creation is the administrator—under God—of discipline, correction, and retributive justice. Thus, in the teaching on the sabbath for the land (Exod. 23, Lev. 25–26), failure to keep God's commandments results in people no longer being supported by the land; violation of God's will results in plagues of creatures (Exod. 8–12).

Implications of This View for Its Conception of God and the Good Life. Thus, in their conception of the environment, human beings should see Creation as good, and before making a judgment against something wrong with the Creation, they should determine whether what they see is due to human failing and sinfulness. People should also respect Creation as teacher—of how Creation is ordered and of what corrective changes in behavior are indicated in the relationship of people with the rest of Creation.

The Nature of the Good Life

The good life is giving your best to the master—glorifying God and enjoying God forever. ("Good" here refers to both its primary meanings: good toward others and in the sight of God and good for oneself.)

Harmonious Living with God and Creation. The nature of the good life is to live in harmony with the will of the Creator and in harmony with the Creation, thereby to glorify God and enjoy God forever.[16] Harmony with Creation means working with other human beings toward harmony and mutual betterment and working in Creation with creatures and ecosystems. Harmony with God means seeking to know God's law, living a life of obedience to God's will, and engaging in prayerful communion with God.

Seeking First the Kingdom of God. The nature of the good life is to seek first the Kingdom of God (Matt. 6:33), not self-fulfillment or self-interest.[17] Fulfillment is a *consequence* of seeking the kingdom. While it is tempting to follow the example of those who accumulate great gain to Creation's detriment, we must work for integrity and harmony in Creation as we pray in word and deed," your kingdom come, your will be done on earth" (Matt. 6:9–10, and "Turn my heart to your statutes and not toward selfish gain" (Ps. 119:36). In seeking the Kingdom of God we must "trust in the Lord and do good; dwell in the land and enjoy safe pasture. . . . those who hope in the Lord will inherit the land" (Ps. 37:3; Matt. 5:5).

Humbly Imaging God's Justice and Love. The nature of the good life is to image God in all things, dispensing justice, loving kindness, walking humbly with God (Micah 6:8).

Humbly Working to Reconcile All Things. The nature of the good life is to reconcile all things, including the uplifting of downtrodden people and other creatures, restoring people, creatures, habitats, and ecosystems. The nature of the good life is not to flaunt our God-given stewardship but to take the form of servants, making peace with all Creation (Phil. 2:5–8).

Observing Sabbath Rests. The nature of the good life is to observe the Sabbath. The nature of the good life is to tend the garden.

Engaging in Confession. The nature of the good life is to confess and repent of attitudes and behavior that devalue Creation and diminish biblical teachings on caring for Creation.

Exercising Forgiveness. The nature of the good life is to forgive. We may have good reason to be upset with those who degrade Creation, but we must recognize them as fellow human beings and potential stewards. Those who have wronged Creation must be given repeated opportunities to share in the work and vision of tending the Garden.[18]

Learning from Word and World. The nature of the good life is to seek to understand what Creation reveals about God's divinity, sustaining presence, and everlasting power, and what Creation teaches us of the God-given order and the principles by which it works.

Authentic Publication. The nature of the good life is to engage in authentic publication.

Implications for Its Conception of God and the Environment. This conception of the good life means that one's understanding of God is as source of all blessing, worthy recipient of praise from all creatures, provider and sustainer to whom gratitude, thanksgiving, and praise are due.

Authentic Publication: Publishing in Life and Landscape

People must not fail to act on what they know to be right. An evangelical perspective on God, the environment, and the good life, is one that puts what is known of God's will for Creation and what is known of the principles and laws by which Creation is ordered and operates into concrete practical form: it must be published in life and landscape. Knowing God's requirements for stewardship is not enough; hearing, discussing, singing, and contemplating God's message in Word and world is not enough; God's requirements must be practiced, or they do absolutely no good. This of course reflects one of the basic evangelical tenets with which we began: the Bible's message of hope and life, of human responsibility for Creation's care and keeping, must not be selfishly kept, but published abroad in land and life.

It is not sufficient to hear the message and express devotion with speaking and singing. If this is the end of the matter, the Scriptures warn that then they are "nothing more than one who sings love songs with a beautiful voice and plays and instrument well, for they hear your words but do not put them into practice" (Ezek. 33:30–32; see also Luke 6:46–49). And again, reinforcing this teaching, Isaiah asks, "Is this the kind of fast I have chosen, only a day for a man to humble himself?. . . . Is not this the kind of fasting I have chosen: to loosen the chains of injustice. . . . to share your food with the hungry and to provide the poor wanderer with shelter" (Isa. 58:5–7). And still again, from James, "Do not merely listen to the word, and so deceive yourselves. Do what it says" (James 1:22; cf. James 1:22–25).

From an evangelical perspective, not only must people publish in life and landscape; they must do so as humble disciples of the Christ. Having been disciples of the first Adam who chose to go his own way, people are part of a lineage that has fallen short of the glory of God (Rom. 3:23). But, affirms scripture, "as in Adam all die, so in the final Adam, Jesus Christ, will all be made alive" (1 Cor. 15:20–22).[19] As disciples of the One "by whom all things were made, and through whom all things hold together," people participate in undoing the work of the first Adam, bringing restoration and reconciliation to *all things*, doing the tasks the first Adam failed to accomplish (1 John and Col. 1; 1 Cor. 15 and Rom. 5; Isa. 43:18–21, Isa. 65, and Col. 1:19–20, 5:17–21). The consequence of this participation in the restoration and reconciliation of all things is authentic publication in land and life: the building of hospitals and ministry to the sick, working for justice and visiting the imprisoned, building schools and colleges and educating the populace, establishing sustainable farms and producing food for society, making churches creation awareness centers and ministering to the environmental needs of the parish, pursuing environmentally redeeming vocations in the world and bringing wholeness to Creation.[20]

Summarizing Conclusion

We have come to an unprecedented knowledge of how the world works (natural science) and how to take care of it (stewardship), and yet there has been no greater degradation of the biosphere in earth's history. The unprecedented environmental degradation we acknowledge and experience today has a single origin: the actions and behavior of human beings. By arrogance, ignorance, greed, or a combination thereof, the whole Creation is being degraded and diminished. We are coming to realize that the legal and technical solutions we have applied toward correcting degrading actions and behavior, while necessary, have not been sufficient. There is a missing element in our response to these degradations: the element of ethics. Knowing how the world works, knowing the principles of environmental stewardship, we now need to develop our understanding of what is right; we need to develop an ethics appropriate to the care and keeping of Creation, appropriate to achieving the Good Life. And this of course, brings us to consider God, the environment, the good life, and their interrelationships.

The good life today, no matter how successfully it may be pursued or achieved, is threatened by the consequences of individual and collective human behavior. Pursuit of the good life may actually be achieving its antithesis. From an evangelical perspective, the ethical standards of the Scriptures provide the best starting point for addressing our human dilemma. Biblical ethical principles, coupled with a good understanding of how the world works, offers promise in achieving the good life for human beings as well as the health and integrity of Creation.

Basic to this perspective is acknowledgment of God as creator and owner of all Creation. The Creator is righteous and just, ordering all Creation consistently according to a comprehensive basic law. Our Creator is prior to and other than Creation, yet intimately involved in it, upholding and sustaining each creature and holding all things in relationships of intricate complexity. God is *transcendent*, while lovingly sustaining each creature; God is *immanent*, while wholly other than Creation and not to be confused with it. God is everlastingly powerful and divine yet enters into personal relationship with people through prayer and the power of the Holy Spirit.

Human beings are made in God's image, and thus are expected to be about imaging God's creative, sustaining, and reconciling love to other human beings, other creatures, and all of Creation. Of all creatures we are the only species able to destroy any other creature and to sustain and care for any other creature, and in this sense we reflect our Creator. But our imaging of God is not to be flaunted or abused but, in the manner of Christ, must take the form of service, even willingness to suffer death. Having been disciples of the first Adam who made the choice for death, we must confess our complicity and participation in the degradation of Creation and follow in the way of the last Adam, through whom the world was made, is held together, and is reconciled.

The environment is a component of Creation, not separable from it, and thus the recipient of this imaging response of human beings is Creation. This imaging response is based on ethical biblical standards coupled with a substantial knowledge of the Creation and the creatures with which we interact. At the core of these biblical standards are three principles: the earth-keeping principle, based on the biblical expectation that people will serve and keep Creation with all of its integral interrelationships (Gen. 2:5); the Sabbath principle, based on the biblical teaching that nothing in Creation, ourselves and the land included, must be relentlessly pressed but rather given times for rest and restoration (Exod. 20, 23; Lev. 25–26); and the fruitfulness principle, based on the biblical teaching that while enjoying the fruits of Creation we must not destroy Creation's capacity for bearing its fruits of vibrant life, clean water, and sustaining substance (Gen. 1:22, 6–9; Ezek. 34:18).

An evangelical perspective requires that knowledge of environmental and Creation ethics, and knowledge of how the world works not be sequestered in books or written statements, but be authentically published in life and landscape. The good news and the good life are to be lived as an inspiration and witness to the others, to the end that others too may be able to find and live the good life in the context of a vibrant, unabused Creation. The good life is one where imagers of God's love for the world come to be able to glorify the Creator and enjoy God forever.[21]

Notes

1. Scriptural quotations in this chapter are from the *New International Version of the Holy Bible* (New York International Bible Society, Grand Rapids, Mich.: Zondervan Bible Publishers, 1978). Used by permission.

2. "*Confessio Belgica*" (The Belgic Confession, 1561, revised 1618) in Philip Schaff, *The Creeds of Christendom, with a History and Critical Notes*, vol. 3, *The Evangelical Protestant Creeds, with Translations* (Grand Rapids: Baker Book House, 1966 [1919]), p. 384.

3. The following are some key references for each of the seven degradations. (1) On alteration of planetary exchange: J. Anderson, D. Toohey, and W. Brune, "Free Radicals within the Antarctic Vortex: The Role of CFCs in Antarctic Ozone Loss," *Science* 251 (1991): 139–46; and B. Bolin et al., eds., *The Greenhouse Effect: Climatic Change and Ecosystems* (New York: John Wiley, 1986); (2) on land degradation: S. Anderson et al., "Estimating Soil Erosion after 100 Years of Cropping on Sanborn Field," *Journal of Soil and Water Conservation* 45 (November–December 1990): 641–44; and T. Plaut, "Urban Expansion and the Loss of Farmland in the U.S.: Implications for the Future," *American Journal of Agricultural*

Economics 62 (August 1980): 537–42; (3) on deforestation and habitat destruction: D. Given, "Conserving Botanical Diversity on a Global Scale," *Annals of Missouri Botanical Gardens* 77 (1990): 48–62; and R. Houghton, "The Global Effects of Tropical Deforestation," *Environmental Science Technology* 24 (1990): 414–21; (4) on species extinction: S. Geer, "One-Fourth of the World's Plant and Animal Species May Face Extinction," *Environmental Conservation* 16 (1989): 372; (5) on water degradation: M. Leistra and J. Boesten, "Pesticide Contamination of Groundwater in Western Europe," *Agriculture, Ecosystems and Environment* 26 (1989): 369–89; J. Maurits la Riviere, "Threats to the World's Water," *Scientific American* (September 1989): 80–94; and D. Moody, "Groundwater Contamination in the U.S., *Journal of Soil and Water Conservation* (March–April 1990): 170–79; (6) on global toxification: E. Atlas and C. Giam, "Global Transport of Organic Pollutants: Ambient Concentrations in the Remote Marine Atmosphere," *Science* 211 (1980): 163–65; P. Larsson et al., "Atmospheric Transport of Persistent Pollutants Governs Uptake by Holarctic Terrestrial Biota, *Environmental Science Technology* 24 (1990): 1599–1601; and K. Reinhardt and D. Wodarg, "Transport of Selected Organochlorine Compounds over the Sea," *Journal of Aerosol Science* 19 (1988): 1251–55; (7) on human and cultural degradation: N. Awa, "Participation and Indigenous Knowledge in Rural Development," *Knowledge* 10 (1989): 304–16; Julie S. Denslow and Christine Padoch, *People of the Tropical Rain Forest* (Berkeley: University of California Press, 1988); S. Gliessman, E. Garcia, and A. Amador, "The Ecological Basis for the Application of Traditional Agricultural Technology in the Management of Tropical Agro-Ecosystems," *Agro-Ecosystems* 7 (1981): 173–85; and M. Oldfield and J. Alcorn, "Conservation of Traditional Agroecosystems," *BioScience* 37 (March 1987): 199–208.

4. See Mike W. Martin, and Roland Schinzinger, *Ethics in Engineering* (New York: McGraw-Hill, 1989).

5. Quotations from *The Assisi Declarations: Messages on Man and Nature from Buddhism, Christianity, Hinduism, Islam and Judaism* (World Wildlife Fund, 1986). Authors are the Venerable Lungrig Namgyal Rinpoche, abbot of Gyuto Tantric University (Buddhism); Father Lanfranco Serrini, minister general of the Franciscan Order (Christian); His Excellency Dr. Karan Singh, president, Hindu Virat Samaj (Hindu); His Excellency Dr. Abdullah Omar Nasseef, secretary general of the Muslim World League (Muslim); and Rabbi Arthur Hertzberg, vice-president of the World Jewish Congress (Jewish). For additional similar material, see C. B. DeWitt, "The Religious Foundations of Ecology," *The Mother Earth Handbook*, ed. Judith Scharff New York: Continuum Publishing Co, (1991), 248–68.

6. Grotius, of course, while giving the Bible a primary emphasis as source material, also derives his ethics and law from philosophers and sages. This could also be done for environmental ethics and law but in this chapter, in view of its title and purpose, I restrict myself to biblical sources.

7. Quotation from P. E. Corbett, introduction to Hugo Grotius, *The Law of War and Peace*, trans. Louise R. Loomis (Roslyn, N.Y.: Walter J. Black, Inc., 1949), xvii.

8. Grotius, *The Law of War and Peace*, 5.

9. This, Grotius says, "has been instilled in us, partly by our reason, partly by unbroken tradition, confirmed by many proofs and miracles attested through every age," for whom it follows "should without exception obey God as our Creator, to whom we owe ourselves and all that we have."

10. Cf. *Oxford English Dictionary*; and W. E. Vine, *An Expository Dictionary of New Testament Words* (Chicago: Moody Press, 1985).

11. *Genesis Rabbah: The Judaic Commentary to the Book of Genesis: A New American Translation, vol. 1, Parashiyyot One through Thirty Three on Genesis 1:1 to 8:14*, trans. Jacob Neusner (Atlanta: Scholars Press, 1985).

12. Psalm 104 is perhaps the preeminent nature psalm in the Bible and should be read in its entirety if the point I am making here is to be fully understood.

13. John 3:16 is perhaps the most widely known verse in the New Testament among Christians: "For God so loved the world that he gave his one and only Son, that whoever believes in him shall not perish but have eternal life." This is followed by verse 17: "For God did not send his Son into the world to condemn the world, but to save the world through him."

14. A passage of note here is Job 40:19: "He is the first of God's works; Only his Maker can draw the sword against him" (quoted from *The Writings: Kethubim* [Philadelphia: The Jewish Publication Society of America, 1982], sec. 3). This I interpret as saying that only God has the authority to kill what God has created.

15. For a detailed treatment of this topic, see Raymond Van Leeuwen, "Christ's Resurrection and the Creation's Vindication," in *The Environment and the Christian: What Can We Learn from the New Testament?* ed. Calvin B. DeWitt (Grand Rapids; Mich.: Baker Book House, 1991), 57–71.

16. "To glorify God and enjoy him forever" is the forumulation given in the Westminster Confession.

17. For a detailed treatment of this topic, see Gordan Zerbe, "The Kingdom of God and Stewardship of Creation," in *The Environment and the Christian: What Can We Learn from the New Testament?* ed. Calvin B. Dewitt (Grand Rapids Mich.: Baker Book House, 1991), 73–92.

18. This is not to say that human beings have the capacity to forgive where only God can forgive.

19. For a detailed treatment of this topic, see Ronald Manahan, "Christ as the Second Adam," in *The Environment and the Christian: What Can We Learn from the New Testament?* ed. Calvin B. DeWitt (Grand Rapids, Mich.: Baker Book House, 1991), 45–56.

20. For putting belief into practice along the lines indicated here, see Art and Jocele Meyer, *Earthkeepers: Environmental Perspectives on Hunger, Poverty and Injustice* (Scottdale, Penn.: Herald Press, 1991).

21. For further elaboration of a Christian perspective on caring for the environment and Creation, see Calvin B. DeWitt, *The Environment and the Christian: What Can We Learn from the New Testament?* (Grand Rapids, Baker Book House, 1991), 57–71; and also Loren Wilkenson et al. *Earthkeeping in the Nineties: Stewardship of Creation* (Grand Rapids, Mich.: Eerdmans 1991), and Meyer, *Earthkeepers.*

6 | JAY MCDANIEL

The Sacred Whole
An Ecumenical Protestant Approach

A religion is a way of organizing life. In our time the dominant religion of the planet is "economism." Its god is endless economic growth, its priests are economists, its missionaries are advertisers, and its church is the mall. In this religion, virtue is called "competition" and sin is called "inefficiency." Salvation comes through shopping alone.

Historically speaking, economism is a relatively new religion. Earlier societies did not have endless economic growth as their central organizing principle but rather ethnic survival or harmony with nature or military conquest or spiritual well-being. Devotion to the god of endless growth seems to have emerged some three centuries ago with the dawn of the industrial revolution in the West. To be sure, this god did not emerge in a mythic vacuum. In some ways, he was the secular son of Yahweh. Like Yahweh, he promised a coming kingdom—a new age—when pain and suffering would cease. Unlike Yahweh, he said, and still says, that it will come about through conquest, not divine grace.

Of course, all religions have their heretics. In our time heretics are those who criticize the ideal of endless growth by suggesting that there are ecological and social limits to growth. In the present situation such heretics are mostly unheard. Imagine a politician who seeks to run for office on a "no-growth" or "slow-growth" platform. He or she would be ridiculed by electorate and media alike, so convinced are many of them—us—that endless growth is the only real hope for human well-being. No growth and slow growth are blasphemous terms. They violate our sense of adventure.

Still, there are heretics in our midst. We find them among environmentalists and human rights advocates such as John B. Cobb Jr., a well-known Protestant theologian, and Herman Daly, an environmental economist. In *For the Common Good: Redirecting the Economy Toward Community, the Environment, and a Sustainable Future* (Boston: Beacon Press, 1989), Cobb and Daly argue that, while service to growth once served its purpose, the costs of growth now outweigh the benefits. They do not say that economic growth is always bad. Particularly for developing nations, growth can be good. But they do say that endless growth is an inadequate ideal for healthy economics or a healthy earth.

Consider, for example, some of the costs of service to growth. Teenagers kill each other for Nike shoes; local communities are torn apart by the movement of capital; rural life is destroyed in the name of urban "development"; animals are reduced to "commodities" on the stock exchange; and the earth itself becomes a stockpile for toxic wastes hazardous to both human and animal health. Amid our devotion to endless growth, so Cobb and Daly argue, the limits of the earth to absorb pollution and renew resources are ignored, along with the need of human beings to live in community with one another, other animals, and the earth.

What is needed, they say, is a sense of balance and maturity. Just as living organisms grow toward maturity and then cease growing, so living economies ought to grow toward maturity and then cease growing. To cease growing is not to die. In a steady-state or mature economy, there would still be a sense of adventure, but the adventure would involve qualitative development over quantitative growth. We would grow in wisdom and compassion, not in gadgets and frills. Moreover, our economy would be integrated into, not wrested from, the larger economy of the earth. We would have arrived at a constant state of material wealth that is comfortable but not excessive, and we would have arrived at a constant state of population that lives within the carrying capacities of bioregions.

Needless to say, few if any economies in our world are mature by these measures. The twin problems of overpopulation and overconsumption imperil us all. We are all addicted to growth. Hence, the need is not for an end to economics or economists but for an end to economism. The need is for economic theories, policies, and institutions that take as their aim not ever-increasing consumption and production but rather human community in an ecologically responsible context. Daly and Cobb go a long way toward spelling out those alternative theories, policies, and institutions. It is no accident that *For the Common Good* won the coveted New Options "Best Political Book of 1989" title. At some deep level, many of us, even amid our devotion to endless growth, seek the balance and maturity of living in harmony with one another and the earth.

Still, proposals such as theirs come to naught unless we ourselves respond to the inner call to community. We must become the kinds of people with the kinds of attitudes that can support ecologically minded and community-based economies. As Wendell Berry made clear in *The Unsettling of America* (San Francisco: Sierra Club Books, 1977) the ecological crisis is not simply a crisis of technology and management, it is a crisis of character. The problem lies not only in the "hardware" of our physical economies but also in the "software" of our minds and spirits.

It is with the need for character, I believe, that the classical religions of the world can play a role. By classical religions I mean corporate paths such as Judaism, Christianity, and Islam; Hinduism, Buddhism, and Sikhism; Confucianism, Taoism, and Shinto. For good or ill, these paths are in the business of character cultivation. Through their creeds, codes, and cults, they provide images of the good life and offer people ways of finding that life. If conversion from the god of growth to the spirit of life is to occur, it must occur with their help and in the context of people who take them seriously.

Here, however, a problem emerges. Heretofore, most of the classical spiritual traditions, with the possible exception of Taoism, have not been particularly good at encouraging environmental awareness and sensitivity. As Thomas Berry has suggested, they have been better at human-human relations and human-divine relations than human-earth relations.

By contrast, the indigenous spiritual traditions—the ways of Native Americans, Aborigines, and Africans, for example—have been more earth-centered. Their way to God has been through the earth. Of course, they themselves have not always been ecologically benign. We need not romanticize their treatments of the Earth to recognize their gifts. What is important is to recognize their gifts. In native traditions, for example, knowledge of one's own bioregion is considered a spiritual discipline in its own right. One cannot travel "the way" without knowing the flora and fauna of one's region, without having what I will later call "a sense of place." By contrast, in classical Christianity, knowledge of one's place has been considered secondary and perhaps even irrelevant to one's salvation. One can be a Christian by "believing in Jesus" and having "faith in God," even if one knows nothing about one's location in nature. When it comes to ecological awareness, Christians today need to become more like Native Americans and less like classical Christians.

My aim in the remainder of this chapter is to suggest one way that Christians might move in this direction. Proceeding from what might be called an "ecumenical Protestant"[1] point of view, I will go on to propose and develop three basic ideas, each of which might help Christians and others become the kinds of people who can help sustain mature economies that are socially just and ecologically sustainable. They are (1) that the

environment is not only an issue among issues but a context for understanding all issues, (2) that, once basic needs are met, the good life is not
one of ever increasing consumption and production but rather a journey
into wholeness in which one knows both "red grace" and "green grace,"
and (3) that God is not necessarily a king or ruler who is "wholly other" to
the universe but rather the "sacred whole" of the universe itself. I begin
with the first idea.

The Environment as Context, Not Issue

Most of us realize that the environment is an issue among issues. We open
our daily newspaper and read about problems of pollution, deforestation,
topsoil erosion, species extinction, animal abuse, and resource depletion.
We speak of the environment as one of our deep concerns, alongside
hunger and violence and racism.

Still, it is also important to recognize that the environment is also a
context for understanding all issues. By "environment" I mean the very
web of life, of which we ourselves are nodes. This means that the environment is not just outside our bodies, it is also inside our skins. Our psyches
and those of other animals are dimensions of the web of life, no less than
the cells composing our bodies and the molecules composing those cells.
Together, we and the other creatures form a context that, in truth, is "our"
context, whether we realize it or not. The web of life is (1) a spiritual
context, (2) a social context, (3) a historical context, and (4) a moral
context. Each deserves explication.

The Environment as Spiritual Context

To illustrate the way in which the environment is a spiritual context, I offer
a story.

Several years ago I participated in a workshop on religion and ecology
for Jewish and Christian seminarians. I was one of several resource leaders
whose task was to help the seminarians better understand the ecological
implications of their own faiths. In the process I realized just how unecological my own point of view could be.

The conference took place at a beautiful conference center in the Hill
Country of Texas. A plate glass window in our meeting room opened out
onto a gorgeous limestone bluff standing over the Frio River. During
breaks all the participants would quickly go outside and stand on a porch,

gazing at the bluff, imbibing its creative powers and nourished by its quiet energy. We could not keep our eyes off it.

A few at the conference had backgrounds in science, and they were able to explain to us that the bluff had a story of its own, written not in words but in lines, textures, and colors. Various sediments of rocks, clearly distinguishable to the naked eye, revealed millions of years of evolution that had preceded and then been shaped by the river. Various kinds of trees—mountain cedar, oaks, sycamores—sat at the water's edge telling stories of the emergence of life in the area, as did indigenous plants shooting out from crevices in the bluff. And various animals could sometimes be seen grazing at the top of the bluff, particularly white-tailed deer, telling stories of individual animals and their various feeding habits during the day.

To be sure, these stories were not verbal stories. Still, with the help of our naturalist guides, they were readable. Nature itself has "writings" that function as "texts" to be read by trained eyes. Gary Snyder puts it this way: "A text is information stored through time. The stratigraphy of rocks, layers of pollen in a swamp, the outward expanding circles in the trunk of a tree, can be seen as texts. The calligraphy of rivers winding back and forth over the land leaving layer upon layer of traces in previous riverbeds is text."[2] During breaks at the conference, we were invited to learn to read such texts.

But one did not have to be a naturalist to be awed by the sheer beauty of the animals and plants and rocks themselves, as they disclosed themselves to us in the here and now. Throughout the conference, we felt God in the earth, not as a person but rather as a power. We felt Holy Wisdom through the sheer presencing—the suchness—of the bluff in its numinous energies. The suchness was itself a story, told not in words but in sheer splendor.

The problem was that during the sessions themselves we did not acknowledge the suchness. When it came to "Christian approaches to ecology" and "Jewish approaches to ecology," the resource leaders referred not to the palpable presence of the bluff but rather to books written in human languages. As we gave our talks to the seminarians, we assumed that our only reliable clues for approaching the earth, our only valid sources of revelation, came from Torah and the Bible. We forgot that the earth itself could be revelatory.

It is not that we said anything untrue. For my part, I spoke of the "goodness" of creation as celebrated in Genesis, of second Isaiah's vision of a peaceable kingdom, of Paul's view that the cosmic Christ is the "pattern that connects" the whole of creation. These ideas are important. They can indeed contribute to ecological sensitivity within Christian circles.

Still, they were ideas gained from a human book. They seemed abstract to the seminarians, most of whom were focused on the presence of the bluff as perceived through the plate glass window. They were all too aware that, in turning to pages of written scripture, I was forgetting the sheer presence of a different kind of scripture, composed of rock rather than ink, with a story of its own. I was forgetting nature as text.

Realizing what was happening, a young student stood up and said: "Turn around. Look out the window. Trust the bluff!" Her point was that I was so fixated upon the written word that I was ignoring the deep feelings that most of us were experiencing in relation to our beautiful setting. We were failing to allow those feelings to be our guides. We were failing to treat our own experiences and the natural world that prompted them as revelatory.

I imagine that most readers already know that the earth is sacred revelation. At least you know this fact when you yourself are troubled. So often, when we need to work through a problem, we take a walk in a park, go outside and sit under a tree, visit one of our "sacred places" in the natural world, spend time with our pets, listen to the wisdom of our own bodies, or gaze at the stars. In so doing we turn to the planet and cosmos—the other 99 percent of creation—for assistance in our deliberations. We hear the voices of nature, not in a way that obstructs our own critical faculties but in a way that nourishes them. Such is the way living scriptures—written or nonwritten—should function.

Still, our religious institutions have not always encouraged us to find guidance in this way. If we are Jews, Muslims, or Christians, our religious institutions often tell us that, in times of trial, we best turn to written scriptures made of ink on paper but not to earthly scriptures made of rocks and plants. It was in the Hill Country of Texas, at the workshop mentioned above, that I realized the inadequacy of such book-centeredness.

In short, the young seminarian was inviting me and others in the room to recognize what was already the case: "the environment" was not simply an issue among issues but also a spiritual context for our own lives.

The Environment as Social Context

In addition to being a spiritual context, the environment—that is, the web of life—is also a social context. By this I mean that the web of life is itself a society to which we belong, whether we realize it or not. By society I mean a "community" of creatures bound together either by kinship, common interests, mutual dependency, or all three.

Under the sway of homocentric ways of thinking, we often think of communities in strictly human terms. When we speak of "members" of the communities to which we belong, we mean human beings alone. In truth, however, the web of life is itself a community to which we belong. Its members include plants and animals as well as people. We are bound together with our nonhuman friends (1) by kinship, insofar as we stem from a common biological heritage: (2) by common interests, insofar as we share with other living beings an eros to survive with satisfaction; and (3) by mutual dependency.

In many ways, of course, we are more dependent on some members of the life community—plants, for example—than are they on us. Life on earth could survive and flourish without humankind. Still, given present population and consumption patterns, much if not most of life on our planet is now dependent on humankind for its survival. In almost all realms save the microbial, other living beings are profoundly affected by our overpopulation and overconsumption. It is not that these other creatures could not live without us. Rather it is that they cannot live with us unless we adopt more benign forms of land management and resource use. Now more than ever, they depend on us even as we depend on them. We form a single life, an ever evolving community.

Process theologians, ranging from the neo Teilhardian perspective of Thomas Berry to the Whiteheadian theology of John Cobb and Herman Daly, point out that this life community is itself a community of subjects and not just a collection of objects. By this they mean that each living being in the web of life has its own interiority, its own inwardness, its own subjectivity. Animals and plants are kin to us, not simply because they share with us a common origin but because they share with us the mystery of subjectivity, be it conscious or nonconscious.

Along with many Buddhists, process theologians suggest further that, directly or indirectly, each subject in the web of life is actually present in every other subject, even as it transcends every other subject. The life community is itself a communion of mutual immanence and mutual transcendence. As the Zen master puts it, "Rocks and trees, hills and rivers, all these are parts of our true self." We are not skin-encapsulated egos, cut off from the world by the boundaries of our skin; we are earth-including selves, enfleshed by other creatures even as they transcend our ego. As the Vietnamese Buddhist Thich Nhat Hanh puts it, we do not exist cut off from the life community, rather we inter-are.

The Environment as Historical Context

As the allusion above to evolutionary kinship suggests, there is still a third way in which the environment is a context as well as an issue. The environment is the historical context of our lives. By this I mean that the environment is not simply a life community but also a *living tradition.* Ordinarily, of course, we think of "traditions" as human alone. We speak of our traditions as the creeds, codes, and cults that we inherit from the human past.

In certain ways, however, these traditions are superficial compared to the still deeper traditions that we inherit from the biological past. Our bodies and genes are the carriers of this deeper tradition. Simple acts of breathing and digesting and excreting are traditions, the ways for which were paved by millions of years of evolution. In truth, the tradition from which we stem originated fifteen billion years ago with a primal flash or big bang. It is an ongoing experiment, full of trial and error, that has unfolded and continues to unfold, galactically and geologically as well as biologically. Our cultural traditions are extensions of the evolutionary process, not exceptions to it. As Thomas Berry makes clear, even our "cultures" are linked with our genes. We are genetically coded to be cultural beings. With every cultural act, we add to a cosmic story that is "our" story, though not ours alone. We are fellow pilgrims with other creatures in a fifteen-billion-year historical process, the future of which is yet to be told.

The Environment as Moral Context

Finally, to say that the environment is a context as well as an issue is to say that the web of life is a moral context. This does not mean that nonhuman members of the web are moral agents responsible for their actions; rather, it means that they are moral patients or moral beneficiaries, deserving our ethical regard. Our "moral community" is not simply the human community; it is the life community.

Of course, there are serious disagreements on the practical implications of this view. In particular there are differences between environmentalists and animal welfare advocates. Many environmentalists, for example, insist that a recognition of the moral considerability of nonhuman creatures properly focuses on ecosystems and species, not on individual creatures. Following Aldo Leopold, arguably the founder of the "land ethics" tradition, they insist that an act is "right" if it respects the "beauty, integrity, and stability" of ecosystems and "wrong" if it does otherwise.

By contrast, animal welfare advocates argue that a recognition of the moral considerability of nonhuman creatures properly focuses on those individual creatures who are our closest biological and spiritual kin, namely, fellow mammals. More specifically, they emphasize our obligations to respect the integrity of creatures whom we have domesticated. They emphasize our responsibilities to free such animals from abuses inflicted on them in factory farms and scientific laboratories.

My own recommendation is that Christians and others recognize the truth of both points of view. The need is to recognize the intrinsic value of individual animals under human dominion as well as the "beauty, integrity, and stability" of ecosystems and then to seek ways of organizing human life that enable animals and ecosystems to flourish, even as human beings flourish. This requires a control of both population and consumption, along with the development of human communities that are humane in their treatment of people and animals and sustainable in their relations with the earth. Such communities can be neighborhoods, local congregations, schools, villages, cities, and perhaps even nation-states. They will be nested within and responsible to the larger context of the bioregions in which they are situated. The Green Party of the United States suggests that human members of such communities strive to embody the following ideals: respect for diversity, nonviolence, community-based economics, grass-roots democracy, political decentralization, postpatriarchal life-styles, responsibility to future generations, personal responsibility, and ecological wisdom. These are the kinds of communities that can support the community-based economics recommended by Cobb and Daly. They are the kinds of communities that move beyond a religion of economism to a religion of life.

The Good Life as a Journey into Wholeness

If humane, sustainable communities are to emerge in our time, we need an image of the "good life" that can support them. The religion of economism tells us that the good life comes through ever increasing consumption. By contrast, the wisdom of the world religions, Christianity included, tells us that the good life lies in subordinating our egos to a higher power or to a deeper self or to the web of life itself. We are told that, if we do so, we will find that true happiness lies not necessarily in getting the things we want but in being free from enslavement to things. Once our basic needs are met, so the religions say, it is time to undertake an adventure of spiritual growth, not ever increasing consumption.

I submit that this adventure in spiritual growth can best be called a journey into wholeness. Wholeness has both an inner and an outer dimension. It includes meaningful relations with the depths of our very selves, and it includes meaningful relations with other people, with plants and animals, and with earth and stars. It comes in degrees, and it is something that we taste but never fully realize in this life. There is no reason the "completion" of the journey cannot occur in life after life. There may well be growth after death.

Wholeness is not the same as moral perfection. To the contrary, wholeness emerges when we realize that we are not morally perfect and when we accept that fact. In Christianity, for example, emphasis is rightly placed on accepting the broken and imperfect dimensions of our lives, cognizant that we are accepted by God amid, not apart from, that brokenness. This is part of the truth of the cross, of what we might call red grace.

In the phrase "red grace," the color red refers to the blood of Christ on the cross and to the color of wine in Communion. For the Christian, the blood and wine are life-giving rather than death-serving because they provide an occasion for recognizing (1) that even God partakes of suffering, as do we and other living beings, and (2) that God does not retaliate with violence for violence. Red grace is the grace that emerges in our lives when we are free to accept our own suffering, realizing that it links us with all other living beings, and when we are free to acknowledge our own greed, hatred, and envy, realizing that we, like all other mortals, are loved by God amid such sinfulness. As Jungians emphasize, only when we can "own our shadows," can we then cease projecting evil onto others at the expense of recognizing it in ourselves.

Only when we can own our shadows can we become healers in a broken world, recognizing that we ourselves are among the first in need of healing. In the current context, our shadows include the idolatry of serving endless growth. Part of our self-awareness and our repentance must lie in finding the god of endless growth within ourselves and then rechanneling his creative energies, not for exploiting others but rather for purposes of earth healing. Understood as our innermost potentialities to sin, inner idols can never be killed. And yet, with the help of red grace, their energies can be transferred from conquest to healing, from death to resurrection.

Still, there is more to wholeness than red grace. In its inner dimensions, wholeness also involves a sense of adventure, through which we feel open to fresh possibilities derived from God for new and hopeful futures; a sense of trust, through which we realize, not that everything will turn out right but that there is a healing spirit, itself divine, at work in our lives and in the world; a sense of inner space, through which we feel centered amid the vicissitudes of life and creative in our capacities to respond to them. A sense of red grace is but one aspect of inner wholeness that, at its best, complements other aspects.

In its outer dimensions, wholeness involves both shared suffering and shared joy. At the level of shared suffering, it involves solidarity with the poor and powerless, the neglected and despised, the forsaken and forgotten. The ecological theologian Sallie McFague rightly stresses that, in our time, the "poor" whom we rightly serve include animals and the earth as well as people. Wholeness involves solidarity with the poor, nonhuman as well as human.

At the level of shared joy, wholeness involves the enjoyment of rich relations with friends and family, colleagues and co-workers, spouses and lovers. In most instances, of course, shared suffering and shared joy go together. Rich relations with other people involves feeling the feelings of others, be they pleasurable or painful. Wholeness lies in empathy.

In its outer dimension, wholeness also involves what we might call "green grace," which is a complement to the red grace stressed above. Here "green" means "environmentally aware." Green grace refers to the healing and wholeness that we find when we enjoy rich relations with plants, animals, and the earth.

The whole-making powers of green grace became clear to me several years ago when I visited a shelter for battered women in Boston. The counselors were deeply influenced by feminist philosophies and theologies, particularly "ecofeminism." I had heard that counselors at the shelter made use of "ecological spirituality" in their therapy, and I wanted to know how they did it. At the shelter I met a counselor who, in a practical way, was an expert on the subject.

For years this counselor had supplemented individual and group counseling with animal-assisted therapy and nature-centered rituals. She would encourage her clients to bond with pets as part of their therapy, to cultivate gardens, to learn about the flora and fauna of nearby parks, to participate in rituals designed to help them know the wonders of the natural world, including the wonders of their own bodies. The results were promising, and she was a firm believer in the healing powers of nature. "The more my clients learn to trust animals and the earth," she said, "the more they begin to trust themselves. And the more they trust themselves, the better they can free themselves from exploitive relationships."

She was also agnostic on the question of God. She trusted the earth and its web of life, but she was not sure she believed in a cosmic spider, a cosmic heart, in whom the web is enfolded. "If grace exists," she said, "it is the grace of rich connections with other nodes in the web. Whether or not it comes from God, I do not know."

She knew that I was a theologian, and she assumed that I believed in a cosmic spider. But she also knew that I understood her reasons for agnosticism and skepticism. I knew that, for her, "God" named a policeman in the sky, a powerful male presence residing somewhere off the planet, whose primary concern was with being worshipped for his own sake. From years of working with battered women, she had had enough of powerful men, human or divine, who were obsessed with being worshipped. They were part of the problem. I didn't blame her for trusting the earth and being skeptical of "God." I sensed that, for her, the Life in whom I believe was itself experienced through the earth and without the use of the word "God."

The life and work of the counselor are instructive in two ways. First, her perspective shows that people can have a spiritual dimension in their lives without believing in God or using God-language, much less being involved in formal religion. This does not mean that God—the cosmic spider—is absent from their lives. For my part, I think God is found in all people, believers or not, and in all living beings in different ways. But it does mean that belief in God can be absent from people's lives, at least at a conscious level. Even as they may not "believe in God," they may nevertheless have a spirituality.

Second, her work reminds us that being rooted in the earth is important, not only because it instills us with attitudes that can help us protect the earth and other creatures but also because it offers us forms of healing we may sorely need. Whether or not we have suffered the pain of sexual abuse, we need rich connections with the web of life and its nodes in order to survive emotionally as well as physically. Like battered women, we need green grace.

What, then, are our opportunities for green grace in relation to the natural world? Consider the following four.

A Sense of Place. The first can be called a sense of place. Most of us already know what it means to have a sense of place, at least with respect to some small portion of the earth. We can recall a natural setting we enjoyed as a child or one to which we return again and again as adults.

In my own case, my earliest sense of place emerged in the Hill Country of Texas, north of Kerrville, on the banks and in the water of Guadalupe River. My parents would take me to the river as a child, and I grew to love the smell of the water, the color of the rocks, the smell of the soil, the beauty of the perch swimming just beneath the surface, the turtles and crayfish, and even the water moccasins, which I feared but respected. If I had to name the spiritual guides of my life, one would be the Guadalupe River. I imagine that you have a guide of your own.

Bioregionalists such as Wendell Berry and Gary Snyder emphasize that we can expand our senses of place to include larger places, the bioregions in which we live. The social worker in Boston indicated one way we might do this. Like her clients, we can learn about the life communities around us. The following test, developed by *Co-Evolution Quarterly*, helped me realize just how much I have to learn about the bioregion in which I live. Perhaps it will do the same for you.

1. Trace the water you drink from precipitation to tap.
2. How many days 'til the moon is full? (Two days slack allowed.)
3. What soil series are you standing on?
4. What was the rainfall in your area last year? (Slack: one inch for every 20 inches).
5. When was the last time a fire burned in your area?
6. What were the primary subsistence techniques of the culture that lived in your area before you?
7. Name five native plants in your region and their season(s) of availability.
8. From what direction do winter storms generally come in your region?
9. Where does your garbage go?
10. How long is the growing season where you live?
11. On what day of the year are the shadows the shortest where you live?
12. When do the deer rut in your region, and when are the young born?
13. Name five grasses in your area. Are any of them native?
14. Name five resident and five migratory birds in your area.
15. What is the land use history of where you live?
16. What primary ecological event/process influenced the land form where you live? (Bonus special: what's the evidence?)
17. What species have become extinct in your area?
18. What are the major plant associations in your region?
19. From where you're reading this, point north.
20. What spring wildflower is consistently among the first to bloom where you live?[3]

The quiz favors country people over urban dwellers, and it favors indigenous peoples over industrial peoples. People in rural areas as in indigenous societies are much more knowledgeable of and attuned to the bio-regions in which they live than are people who live in cities.

Reverence for Life. A second way in which we can be rooted in the earth was also illustrated by the social worker in Boston. Recall that she encouraged her clients to develop close relations with animals, specifically pets.

Close relations of this sort involve knowing animals as living subjects with value in their own right, as opposed to mere objects of value only to others. This kind of knowledge is at the heart of the animal rights and animal protection movements. Following Albert Schweitzer, we can call this knowledge "reverence for life."

Amid reverence for life, the focus is not on identifiable geographical regions but rather on individual kindred creatures, particularly animals, who are our closest psychological and biological kin. We have reverence for life when we feel kinship with their joys and sufferings and when we want for them the kind of happiness that they want for themselves.

Reverence for life is an antidote to points of view that emphasize systems but not individuals. Sometimes the sense of place described above can lapse into such insensitivity. It can celebrate the web of life as embodied in a local bioregion but forget the nodes in the web. This is like loving forests but neglecting individual trees or like loving humanity but hating individual people. To avoid such abstractness, a sense of place needs to be complemented by reverence for life. The "land ethic" of Aldo Leopold needs to be complemented by the "life ethic" of Albert Schweitzer.

Among the religious traditions of the world, the ones that have been most keenly reverential of life are not the indigenous traditions, important as they are. Rather they are the classical traditions of Jainism and Jain-influenced Buddhism, with their doctrines of *ahiṃsā*, or noninjury to animals. As these Asian traditions make clear, the life of compassion rightly extends to animals as well as to people. It rightly leads to a progressive disengagement from injury to animals.

Reverence for the Planet. A third way of being rooted in the earth lies in feeling a sense of identity with and reverence for the planet Earth as a whole. Such a feeling was implicit in the mind of the social worker when she spoke of "trusting the earth." Here the planet functioned for her as a subject of loyalty in its own right.

In a certain sense, loyalty to the planet is a new possibility in human history. Many people throughout recorded history have had a special sense for the bioregions in which they lived, but few were able to identify with the planet as a whole because they had no way of seeing or visualizing the entire planet in its cosmic context.

For us the picture of the earth from space has made such visualization possible and easy. When mention is made of the earth, most of us now imagine a beautiful globe cast in relief against the stars. The earth as a whole has become a mythic image in our imaginations. It thus provides food for new mythical sensitivity.

Part of this mythical sensitivity involves seeing the planet as alive. This can mean several things. Some imagine the planet as a living object in its own right: having awareness of its own, not unlike the way an animal has such awareness. Still others speak of the Earth as a living subject but with awareness more diffuse and less centralized than that of, say, a cat or dog. As they see it, the earth's subjectivity is like that of a living tissue or a complex plant, rather than like that of an animal. And still others (and I count myself among them) speak of the Earth not as a subject in its own right but rather as a community of subjects, like a forest whose "spirit" is the sum total of the spirits of each of its living beings. In each of these instances the earth itself becomes a subject of reverence and loyalty. When we are loyal to the earth, we are loyal to something that has an identity of its own, of which we are a part but which is more than we. We feel connected to and part of a larger whole—our planet—which is itself connected to a still larger whole, namely, the cosmos. The natural extension of loyalty to the earth is cosmic awe.

Awareness of Our Bodies.   A fourth way of being rooted in the earth is closer to home than the loyalty to the earth. It lies in being aware of our own bodies as living incarnations of the earth's energies. Our very closest contact with the earth comes not in our knowledge of our bioregion nor in our allegiance to the planet, nor even in our sensitivity to creatures around us. It comes through simple acts of breathing and eating and walking and sleeping. Each of these acts is an instance of the living dynamics of the earth. As is emphasized in many forms of Buddhist meditation, breathing itself is the self-awakening of the cosmos. We need go no further than our own breathing to experience enlightenment.

Of course, many of us are not very aware of our bodies. Unfortunately, in the West the body often has been considered relatively unimportant in spiritual pursuits. It has even been treated as an enemy to be transcended. Some believe that the more spiritual they are, the less embodied they will be.

On the other hand, some of us can be aware of our bodies only as objects that we wish conformed to social standards of beauty. We are aware of our bodies only as others see them, as commodities for their consumption. We wish that we were "prettier" or "more virile looking" or that we had less gray hair. We want to look like the models we see on television. This is not the kind of body awareness I have in mind. I have in mind instead an immediate awareness of the body as subject of trust and as source of wisdom.

To trust our bodies is to realize that they are the accumulated wisdom of millions, indeed billions, of years of cosmic, geological, and biological evolution. The history of the cosmos has, in its own way, been a process of trials and errors, amid which things have been learned, such as how to regulate temperature and how to heal wounds and even how to think and feel. We do these things by virtue of our genes, which carry within them the memories and wisdom of the distant past. To be rooted in our bodies is to recognize that our bodies can be spiritual guides, teachers. Like all teachers, they are finite. We cannot learn all that we need to know by listening to our bodies. But we need to listen to our bodies even as we also learn from other sources. They may well carry dreams, revelations, that are the voice of God itself as channeled through evolution.

God as the Sacred Whole

In the section above, I identified four dimensions of a healthy, whole-making spirituality, four dimensions of green grace. These aspects of eco-spirituality can be internalized by the religious and nonreligious alike and by people who believe in God and those who do not.

In my own case, I internalize them through a belief in God. I understand them to be four ways in which I feel the presence of a Life within and yet more than the cosmos itself. I believe that this Life was shown distinctively, but not exclusively, in a carpenter from Nazareth two thousand years ago. In him, but not him alone, I see the cosmic spider in whom the web of life is enfolded. I believe his death on the cross is itself an occasion for experiencing red grace.

But I well recognize that people can enjoy these four ways without believing in the Life, much less in the carpenter. One does not have to be Christian in order to find God. One of our best hopes, if the religion of economism is to be transcended, is that Jews in their ways and Muslims in theirs and Buddhists in theirs and Free Spirits in theirs can come to enjoy the healing that comes from a sense of place, from reverence for life, from loyalty to the planet, and from trust in the body. Only as we learn to enjoy green grace can we ourselves become healers of a broken world.

Still, Christians today are in need of meaningful ways of envisioning the God in whom they believe, who was revealed in the carpenter. I bring this chapter to a close by suggesting one image in which God can be envisioned. Following Cobb and Daly in *For the Common Good*, I suggest that God can be envisioned not as a king or ruler external to the universe but as the sacred whole of the universe itself.

Ecological theologians such as Sallie McFague are deeply critical of the monarchical view of God. This is the image of God as a being among beings who resides off the planet and whose will is that we serve him much as plebeians serve their kings. McFague criticizes this way of thinking about God for at least three reasons: (1) it inhibits human fulfillment by leaving people in a position of perpetual adolescence in relation to God, (2) it lends itself to the assumption that God is concerned with human beings alone, and (3) it seems to externalize God so completely that God seems disconnected from the universe save through "his" one incarnation in Jesus. Elsewhere I have argued that the monarchical image can, in some circumstances, be more useful than McFague recognizes.

Here, however, I want to grant some truth to McFague's point in order to propose another image which, in my view, can help Christians enter more deeply into ecological sensitivity. It is the image of God as sacred whole.

There are, of course, many different kinds of wholes, some of which are and some of which are not more than the sum of their parts. In the case of the divine reality, I wish to speak of a whole that is greater than the sum of its parts, but the question emerges: In what way? Let me offer two options.

One is to say that God is more than the totality of beings in the universe in the same way that a city of twenty-five thousand people is more than the totality of people. A city is more than its people in the sense that its dynamics or governing principles are not reducible to the psychological dynamics of its individual inhabitants, but not in the sense that it has a psychology of its own. Whereas the various inhabitants are subjects of their own lives, we do not ordinarily imagine a city as the subject of its life. We do not recognize cities as having agency or consciousness. Accordingly, to say that God is more than the totality of beings in the universe in the way that a city is more than its inhabitants would be to say that there are laws of the universe that transcend all particulars but not that the universe itself, considered in its unity, has agency or consciousness.

A second option, however, is indeed to imagine the whole as having agency or consciousness. This would be to envision God not on the analogy of a city or its inhabitants but rather on the analogy of a mind or soul or spirit to its body. From this vantage point, the sacred whole would be a life, with agency and consciousness of its own, not unlike the way that you the reader are a life with agency and consciousness of your own. Just as your own body is gathered into the felt unity of your own life, so the universe is gathered into the felt unity of the divine life.

This second option is the way of imagining God that I wish to commend to Christians and others. It is to suggest that the universe itself is the very body of God and that the ultimate unity of the universe is itself a life. This has at least three implications.

First, it suggests that, just as what happens in our bodies happens in and to us, so, according to this image, what happens in and to the universe happens in and to God. This means that God shares in the joys and sufferings of each living being on our planet, much as we share in the joys and sufferings of our own bodies. Such is the image of the cross in Christianity. It presents the image of a God who suffers and is vulnerable. My proposal is that this image of the cross be taken quite literally. Wherever there is suffering—be it in the pain of a hungry child or the terror of a hunted deer—there is divine suffering; and wherever there is joy—be it in the delight of a child at play or the frolicking of a colt in pasture—there is divine joy. To say that God shares in the joys and sufferings is to say that at the very core of the universe there lies deep, abiding empathy. When we ourselves partake of such empathy, we participate in the very reality of God.

Second, just as some things happen in our own bodies that we cannot prevent, so, according to the image, some things happen in the universe that even God cannot prevent. Just as cells in our bodies have creative capacities, for good and ill, that transcend our wills, so cells in God's body—including human beings but also porpoises and microbes—have creative capacities that transcend God. The sacred whole of the universe cannot and does not control the unfoldings of the world like a puppeteer; rather it beckons each creature, given its unique nature, to find that wholeness, moment to moment, that is possible in the situation at hand. The various forms of communion, differentiation, and subjectivity that we find in the universe as a whole, as well as the eros within each creature to find its individual wholeness, are evidence of divine influence but not divine coercion. From the vantage point of the sacred whole image, the creatures of the universe experience the divine as a calling presence whose aims require their response for realization.

Third, just as our primary aim for our own bodies is healing and wholeness, so the aim of the sacred whole for its body is healing and wholeness. The good life, as envisioned above, involving a sense of both red grace and green grace, is a distinctively human form of wholeness. It is what God wants for us and we want for ourselves. Other creatures have their forms as well. The sacred whole is within each creature, not as a manipulator of its life but rather as an inspiration to its own creativity. Sometimes, perhaps often, we fail to hear the call of its guidance. Such is the reality of human sin. Still, the sacred whole calls us and all creatures into wholeness. The defining characteristic of the sacred whole is unconditional and unfailing love.

Clearly, the sacred whole is more than the universe but not in a way that makes it a puppeteer or king. It is more in the sense of being an enveloping context in which, to quote Paul in the New Testament, we "live and move and have our being." We never experience the whole as something outside that we can objectify from external perspective; rather we experience ourselves as within the sacred whole, within God.

Understood in this way, the sacred whole is itself the ultimate environment of our lives. It is the Environment with a capital E, within which all environments with lowercase e's are situated. My suggestion in this chapter is that these two kinds of environments—God and the web of life—are deeply connected. It is also that the good life, for us, involves openness to both and to the many ways in which they are connected. We know God-the-Environment, and we know our own environments, through a sense of grace both red and green. Should we enter into this grace more deeply, we will move beyond the religion of economism into a spirituality of life. In the long run and even in the short run, such spirituality is our only true hope. Only with such a spirituality can we become the kinds of people with the kinds of attitudes that can support the humane and sustainable communities we sorely need.

Notes

1. Ecumenical Protestants are also called "liberal Protestants." Standing in contrast to evangelical Protestants, whose primary authority is the Bible and for whom dialogue with other religions is not a high priority, ecumenical Protestants are those (1) whose sources of religious authority are reason and experience as well as scripture and tradition, (2) who approach these sources as dialogue partners but not absolute authorities, (3) who are open to the distinctiveness of Christian truths but also to the distinctiveness of other religious truths, (4) who reject Christian exclusivism, and (5) who believe that Christianity itself is an ongoing social movement capable of growth and change.

The chief strength of ecumenical Protestantism is its openness to truth, goodness, and beauty wherever they are found. The chief weakness is its tendency to be too open, at the expense of firm rootedness in the best of the Christian heritage, and too easily co-opted by the religion of economism. If the "sin" of evangelicals is bibliolatry, that of ecumenicals is lukewarmness. Accordingly, ecumenical Protestants and evangelical Protestants need one another. A healthy Protestantism needs both commitment and openness, roots and wings. Evangelicals remind ecumenicals of the need for roots: ecumenicals remind evangelicals of the need for wings.

In what specific communions or denominations do ecumenical Protestants find themselves? Some are Methodist and some Baptist, some Episcopalian and some Unitarian, some Lutheran and some Presbyterian. Some are even Catholic, while embodying the spirit of ecumenical Protestantism. In the house of ecumenical Protestantism, there are many mansions. In any case, increasing numbers of ecumenical Protestants now seek to widen the very notion of ecumenism. Whereas earlier ecumenism named a desire to be in dialogue with other communions within Christianity and with other religions, it now comes to name a desire to be in dialogue with the earth. Ecumenical Protestants recognize that historical Christianity has been unecological, and they hope for the emergence of a more ecologically minded Christianity.

2. Gary Snyder, *The Practice of the Wild* (Berkeley, Calif: North Point Press, 1990), 66.

3. *Co-Evolution Quarterly* 32 (winter 1981–82):1.

7 | ALBERT J. FRITSCH, S.J.

A Catholic Approach

Theology, or "faith seeking understanding" as St. Anselm defined it, takes us on a historical journey in the sequential covenant relationships of God and the believing community. Truly, this is a challenge for those with formal theological education. But isn't the defense of environment part of understanding our shared faith in the value of this earth and the God-given vocation to heal its wounds and thus bring on the fulfillment of the good life?

A Catholic perspective on God, the environment, and the good life has a basic core understanding and invites diversity of expression. In this brief treatment it is difficult to discuss with any thoroughness both this core understanding and difference. Needless to say, to overemphasize either one would be a mistake. I prefer to follow a more process-oriented approach to these subjects in the manner that fellow Jesuit, Bob Sears, and I are taking in our forthcoming book: *Earth Healing: A Resurrection-Centered Spirituality*. Bob's previous reflections dealt with the five levels of faith in God; my *Earth Healing Guide*, which is being written, will extend the ecological discussion to include deepening and more conscious levels of concrete applications to physical facilities, edible landscaping, woodlands, water, wildlife and wildscape, conservation and renewable energy, food preparation and preservation, waste management, transportation, and indoor environmental resources. I hope this assessment process will initiate a good life that is also green.[1]

Divine Life: The Blessing of Our Creating God

"See, I am establishing my covenant with you and your descendants after you and with every living creature that was with you: all the birds, and the various tame and wild animals that were with you and came out of the ark" (Gen. 9:9–10).

Our Catholic tradition proclaims a personal, loving, creating God, who calls each of us by name and who invites us to be part of the divine family. This God is one, eternal, and all-powerful, both transcendent and immanent, always caring and provident, and most willing to answer our prayers and petitions. We believe that God's Chosen One has entered into human history, being born of the Virgin Mary, living among us, working miracles and wondrous signs, demonstrating God's love through compassion for the poor, confronting the power structures and gaining their animosity, and accepting suffering and the cruel death of the cross for our salvation. After Jesus is raised from the dead he is taken up to Heaven, and the Holy Spirit comes down upon the assembled church and remains with her until the end of time, enlivening her action and encouraging all believers to live the trinitarian life.

At the very heart of our Catholic belief is the triune God—creating God, redeeming God, and enlivening God—one God and three persons, a mystery beyond us and enveloping us. Thus we sign ourselves with a cross in the name of the three persons of the Trinity and we profess in so doing that God's image and activity is found in all creation, both in the pristine states of nature and also among the suffering and wounded portions of our earth. Insofar as we are the ones who wound our earth, we are called by our loving God to become co-creators, co-redeemers, and co-enliveners. In fact, our psychological processes, our way of acquiring knowledge and daily living, and the manner in which our communities are formed and function have the trinitarian mark—as do all natural processes, especially that of earth healing itself. While untouched earth bears witness to the Trinity, all the more does suffering earth now being healed.

We respond to this majestic divine call to action through praise, thanksgiving, and celebration of our work, for liturgy is "the work of the people." Within this work we manifest our belief in the mystery of the Resurrection by bringing back to life this tired and wounded earth and by helping raise what was wonderfully created into a more wonderful re-creation. The life of the community of believers is enhanced through participation in divine worship and sacraments. The liturgy of Word and Eucharist is a transforming and empowering event. It is no accident that our Western technology developed among a people steeped in the Eucharist. Rather it is a natural progression of people learning to create and demanding the time it takes for prayer and reflection. Since all believers are called to undertake this task, there is little room for elites of any sort. All people need free time to participate in the creative activity, and thus technological progress is ultimately a liberation of all people from the pure drudgery of subsistence work.[2]

In a spirit of eucharistic thanksgiving we accept this world as God's gift, which beckons gratitude found in service to all creatures as Jesus has served us. Through abnegation and fasting we establish our own inner harmony and ecological balance for gaining self-control so that we act more in the spirit of the humble Jesus. We join with others and gradually grow and develop and become one with a community of believers in sharing in blessings of every kind, ranging from the delicate nature of earth to our health, senses, friendships, and days of life. As co-creators we become more acutely aware of our present time, place, and neighbors, lest we become disoriented and spend precious time regaining our bearings. That means we know

- the locality and biological community or bioregion in which we live, the "here";
- the time of day and night, the climate, how the wind blows, and the change of the seasons, the "now";
- the neighborhood and the many people who enter and make our life meaningful, the "we."

Awareness of the "here," "now," and "we" throws us into the hands of our Creating God and allows us a chance to understand more fully the ecological principle of the interconnectedness and interrelationship of all creatures.[3]

Environment: The Making Whole Again through Our Redeeming God

A community's growth in understanding and awareness of environment is analogous to its growth in faith. The fledgling ventures out cautiously, reaches out to others, falls, recognizes the need for assistance, and moves far more securely with a companion. Maybe that venture of childhood exploration is repeated in many and diverse ways throughout one's life. The Catholic perspective invites love of, understanding and compassion for, and an active participation with others in defense and bettering of the environment around us. But that involves learned experience, which takes time and effort. To grow in an understanding of environment follows patterns similar to our growth in faith and awareness of others, especially the poor. This pattern is worth mentioning here.

Stage One. We first encounter our environment through the eyes of one who sees all things as totally new and enriching and readily responds enthusiastically. This is the experience of environment as blessing; it includes plants and animals and all beings that can instruct us by their presence. These creature-teachers fill us with exuberance, energy, laughter, and song, and they invite us to extend our area of concern to those who do not necessarily share our belief systems or culture. It is the poetry of Francis coming alive. If not eroded by life's troubles, this experience of creation continues into adult years when encountering a sunrise, a beautiful natural site or tree or animal, or the quiet star-filled sky at night. We realize this in the widespread popularity of ecospiritualities, in outdoor exercise and activity, in travel to exotic places on so-called eco-tours, and in the enjoyment of geographic narratives and photography of wilderness areas. Environment in this sense captivates and elevates the human heart, but people can idolize it and not go deeper, feeling content to remain at this level, even naming this experience of nature "God" or "Gaia."[4]

Stage Two. A deeper reflection makes us aware that mere seeing is not sufficient. If we treasure what we see, we are drawn to act and assist others, even if the actions we perform are imperfect. By acting we experience power surging within us, and we are tempted to control, dominate, and subjugate the less powerful. We stand at the great moment between hunter/gatherer and cultivator. A plowed field may be regarded either as an extension of culture and civilization by agrarians or an abomination by ardent wilderness lovers. Is reshaping the natural bad or good, artificial or all the more natural? Few will refuse the produce of those fields just because they are cultivated fields even when they question the process of extending cultivation to more and more wilderness. Environment at this stage is not just beautiful but also resourceful, productive, and able to be shared with others and controlled by some. Its shadow side is self-evident to all here present.

Stage Three. The drumbeat of Genesis telling us that all of God's creation is good is muffled by the sin of Adam and Eve—the presence of evil in our midst even in a garden of goodness. That sin manifests itself today in both individual and social disorder, in the breakdown of families and domestic violence, and in vast and spreading social and environmental disorder—in acid rain, demolished forests, chemically addicted farmland, overly developed travel corridors, trashed and littered landscape, and the festering hazardous waste dumps. Environment turns from joyful to suffering creation, and we need to feel it.

This is why the country is in mourning, and all who live in it pine away, even the wild animals and the birds of heaven; the fish of the sea themselves are perishing. (Hos. 4:3)

Wasted lie the fields, the fallow is in mourning. For the corn has been laid waste, the wine fails, the fresh oil dries up. (Joel 1:10)

Human insensitivity and greed and the silence of those who could speak but were afraid all add up to a sorry lot called sinful us. We simply repeat Shakespeare's words in *Julius Caesar*, "Pardon me thou bleeding piece of earth that I have been so silent with these butchers." To even admit our own participation in the butchery is a moment of powerlessness and nakedness, of darkness of the soul. However, through faith we know it can be a grace-laden event.

Stage Four. Responding to the frail nature and wounded condition of our environment means realizing what we have individually and collectively done wrong. In wounding we become wounded. In wanting to heal we want to be healed through God's good grace. Hesitantly, we seek forgiveness of both Creator and wounded creatures. Environment is no longer distant even as suffering; it is ever closer at hand, and our growing compassion brings us companionship from co-sufferers. No prayer goes unanswered, and that is ultimately why we pray. As divine forgiveness comes, and surely it does, then new power surges. Arise and walk. Suddenly, painful process in utter desolation becomes a moment of consoling light, resurrection, new hope of victory.

With confession of wrongdoing comes the demand for restitution for the earth that has been damaged. Faith takes on a dual aspect of realizing that we are forgiven and realizing that we need reestablish justice. It is not merely feeling good about helping others that draws us to activism. Rather it is a question of *justice*, a justice rooted in biblical tradition and well worth dusting off and contemplating. A lasting environmental activism is not nourished by a blissful and innocent enthusiasm for creation, but because we have damaged the seamless web of life, we need to restore it. We are not onlookers or brash activists unleashing the power of new-found technology; rather we have been humbled and understand that through forgiveness and with God's grace we can refashion something better. In healing earth, earth heals us; in making earth new we are renewed in and through earth.

Stage Five. Ever deepening levels of consciousness take us to where we identify with others, where distance is reduced and where the "we" includes a community of all beings. This more mystical level is not easily attained but is clearly ahead of us and is an invitation to enter at the divine invitation. Those experienced in spiritual journeys tell us there are no shortcuts, no set pattern that can automatically trigger such levels, no literacy programs or college courses, no swirling images and cosmic fantasies, no magic. Nor do we want to make this so esoteric that it excludes a greater part of the human race. Yet we realize that in the modern materialistic culture the evil one tempts us to be consumed by consumer goods and so distracted that this deeper fifth—or even second through fourth—stage of identification appears remote and alien.[5]

A Catholic perspective has always accepted the reality of pilgrimage, which must include our eco-Emmaus journey. Were we so blinded that we forgot that even the earth had to suffer so as to enter into its glory? And does not this require prayerful discernment of good and evil. Spiritual life and growth accepts mistakes. Environmental messianic vocation is not the glory road; it is quite dusty out there. What holds us back as eco-pilgrims is too much blame, too little shame. The earth calls for more than a first-stage creation centeredness. It cries out in anguish for a fundamentally sound and heart-penetrating redemption theology that does not excessively dwell on sin but proclaims forgiveness and the need for restitution, an action that prepares us to become a resurrection-centered people. I agree with Wes Jackson, who says we need to spend some time reexamining and developing our redemption theology.

The church stands apart from any fundamentalism, whether referring to squabbles within church circles or the more recent literal interpretation of Scripture or, for that matter, any form of earth fundamentalism that imposes theological interpretations on geological or cosmological data and theories. Any fundamentalism has the elements of obtaining an insight in its entirety fairly quickly and does not admit of changes of that insight; and the primary efforts of the believer are to guard the repository of truth and to spread word of how a personal experience or insight has affected belief through being born again in the earth. The weakness of many forms of fundamentalism is the neglect of justice—and we cannot possibly heal this earth no matter how lofty the language unless we make restitution for our wounded earth. The church's quest for this justice is part of its mission, and with something as grand as the earth itself, we need to muster the resources of all people of goodwill.

Throughout the two millennia of the church's existence there has been growth in consciousness of what justice means and how it is to be extended to people and to other creatures as well. That struggle of making whole or integral again does not cease. Even the art of healing improves with the years, and we all learn from past conflicts, which are seldom resolved through capitulation of one party but rather through mutual understanding or complementarity. The outmoded polarity of creation-centered versus redemption-centered spiritualities needs to give way to the completion of redemption through re-creation, or a resurrection-centered approach, which is more in keeping with the Catholic perspective.

Just as any reflection on our creating God leads us to the ecological principle of interrelatedness, so in reflecting on redemption we discover the second major ecological principle: nature does not waste—all is recycled and reused and serves as resource for a new process. Wastes are not nouns created by natural process; they are the products of wasteful people. Through our redeeming God, that which was wasted through our human wrongdoing is now saved and made whole again.

Good Life: Our Enlivening God

The messianic promise of a new creation brings us to what Catholics mean by the "good life." This is a life that goes on beyond our mortal existence; it is life after death that is already being partly realized here and now—and that is worth celebrating and even toasting and dancing and singing. Why not? We are a people in the process of being saved.

The Catholic notion of the good life is based on
- the worthy aspiration to celebrate with all creation;
- an understanding that creation is limited and an extending consciousness that resources are not equally shared;
- a call to be moderate in resource use and in the control of our appetites, and that may now be extending to family size;

- a growing awareness that by using our talents we give health and well-being to those around us and become more healthy in the process.

The suffering, the imprisoned, the jobless, the endangered species, all of these cry out for a good life, and all our actions need be pro-life. In response, our action that leads to good life, whether it be the bone-tiredness in helping flood victims, sharing of one's meal with the stranger, or talking when someone needs a companion, is the dawn of good living. The good life involves reestablishing eco-justice, not attaining personal happiness by enjoying a pristine and vanishing environment. Ironically, in bringing about justice we do find peace and divine consolation, though we sometimes neglect to appreciate it as God-given and grace-laden.

The good life is not a mirage, an unfulfilled dream, a future divorced from the present, a preoccupation with death culture, an illusion of the evil spirit, a sad joke of the cynic. Nor is it

- the drive for personal pleasure, achievement, and contentment;
- the quest for luxury or superabundance;
- the many forms of segregation that become a hell on earth;
- the license to use up material resources at one's will;
- basking in fame, good fortune, or the security of the remaining years;
- a survivalist or bunker mentality that protects one from the influences of the pervasive culture.

The good life challenges us to conserve and protect and monitor resource use. If we consume resources that could have been used to produce basic necessities for our fellow human beings, or if we squander resources that could have made life more livable for plants and animals, then we live the bad life, no matter how luxurious the life-style or consumer-ridden the culture. Rather the good life is taking care of the needy, being sensitive to their anticipated necessities, and being willing to halt those who take unjustly from the earthly commons. The good life includes giving consolation and hope, forgiving those who desecrated land, and offering appropriate alternatives for reestablishing order where chaos reigns. The good life is being moderate in all things and recognizing their value in their thoughtful use.

The good life is to follow Jesus to Calvary and beyond the grave. The good life is the resurrected life, the firm conviction that Jesus rose from the grave and that we in turn in our belief in this resurrection event will help enliven this suffering earth. The good life is Eden, which is being reestablished here and now in the quiet places of our land, especially among the women religious, who seem to know instinctively the healing arts so much better than men do.[6]

The good life commits us to an ever deepening spirituality that utilizes all of the individual talents that each of us has and the encouragement of others to develop their unique ecological niche in the world in which they find themselves. Variation and differences are most important, and much depends on just how free we are to express our inclinations to action in different ways, such as by respecting our neighbor's rights and person (fellow creatures), conquering ourselves as masters of responsible means of preserving and using (masters), caring for what has been of such temporary stewardship (stewards), compassionate regard when harm is done (good Samaritans), defending what is harmed through active monitoring and service (suffering servants), eagerly learning and revealing the mysteries of the earth (educators and students), being willing to stand up and denounce the injustice done to others (prophets), and lastly, celebrating and laughing with our smiling God at our own incongruities (comics). If blessed with a long life, we may even become wise and manifest wisdom as part of the good life.

The earth's people need to expand their tents and vision. The task is too great. All people of goodwill should be invited to participate in the global enterprise. And this expansion includes coming to terms with the good life as aspired to by simple people, namely, sufficient food, a safe and weatherproof place to call one's home, good roads for travel and supplies, enough to wear, proper education, a healthy environment, and a decent place to recreate and celebrate.[7]

These are the aspirations of the poor, those who have yet to taste the fullness of justice. To ignore those simple aspirations is to fail to grasp the good life as seen and articulated by people of all races and faiths, who seek life in a fuller way. Granted, this justice is anthropocentric in its conceptualization and its implementation. If we do not hear needy people when they call, we may not hear the more subtle earth cry that requires sensitivity and deep listening. On the other hand, to say we are biocentric or geocentric and ignore human justice issues is to create false dichotomies and allow for the perpetuation of a world of haves and have-nots ever more ready to break out in armed violence.[8] Prosperity of all in the community of being is essential for the good life and the ultimate security of all the earth's inhabitants.

Focusing is difficult, but excessiveness of any type is contrary to the Christian message. It blinds one to insufficiency on the part of our neighbor and to detours into preserving what is fleeting, of less worth. For the believer the preferential option is to see, to stop, to treat immediately, and to look after until healing occurs. We certainly have trying times and new circumstances, for never before did people know they could destroy—or save—the earth in the clarity that we do. But we also have greater dreams, hopes of victory, a clarity that says we can come to a deeper level of spirituality if we but move forward.

Focusing according to our individual talents affords opportunities for diversity in expression and fruitful results. The Catholic perspective calls all to use their talents to address the environmental crisis and to move forward to the good life. In fact, the 1991 bishops' pastoral letter "Renewing the Earth" calls on scientists, educators, parents, theologians, business leaders and workers, members of the church, environmental advocates, policy-makers, and citizens to participate in the process of shaping a nation and world more committed to the universal common good and to an ethic of environmental solidarity. The third major ecological principle—of greater variation as ecologically healthy—proceeds from that of creativity and redemptive activity discovered in our interconnectedness and the need to conserve resources. Our enlivening God invites the many talents of the human family to be brought forth, all people finding their respective niches in the building up of the New Heaven and the New Earth.

Conclusion

We have seen three aspects of growth. First there is the manifestation of our creating God always at work in our world and never ceasing to make anew our tarnished activities. In the second section we see that part of the process in which we are invited to join is saving or healing the earth, which is redeemed in the blood of Christ and the call to make restitution for wrongdoing. The third section deals with the work of our enlivening and ever renewing God, responding to the cry of the poor expressed in our own individual ways. The new Eden has begun, and the good life is beginning to be lived and within the reach of all people. It is not the privilege of the elite or the chosen few. Three in one, a Trinity at work—this is the good news.

Notes

1. This *Earth Healing Guide* began as a compilation of explanatory notes that are added to our sixty-five audits or assessments of nonprofit organizations. The written reports offer

a systems approach to more ecologically harmonious ways of using specific community resources and at the conclusion sketches a ten-year plan for action based on the group's human expertise and physical resources. What is becoming evident in constructing this guide is that the very manner in which one initiates and conducts an ecological design is itself an ever deepening spiritual journey that admits of stages of growth. The beauty is not some artificial termination or product but the commitment to be on an ecological journey of service to others—human and nonhuman creatures.

2. A fuller description of some of these ideas about technology and Christian belief is found both in *Renew the Face of the Earth* (Chicago: Loyola University Press, 1987) and in my forthcoming contribution to the Catholics and the Environment series that is being edited by some of the people associated with the University of New Hampshire.

3. This principle needs continual application so that we begin to understand fully how we are interrelated. I recently attended our Ohio Valley Bioregional Congress and was struck by how many spoke in cosmic, global, and interpersonal terms but neglected to mention the existing site, which was filled with Franciscan warmth and peace. This coming to know our feelings, this association with the good spirit, is at the heart of any true sense of "home" and any orientation within our world. I have emphasized this point in *Down to Earth Spirituality* (Kansas City, Mo.: Sheed & Ward, 1992) and will continue to probe the mystery contained in being truly here, now, and we. One shouldn't leave a gathering without touching the soil, tasting local produce, sensing the weather, and feeling the quest of those present.

4. I use "eco-tour" in its now popularized manner of sightseeing with an ecological and educational purpose. It does not mean I fully endorse this practice; however, it is a way for the affluent to break out of their limited world of distracting concerns and be open to deeper calling. The eco-tours we sponsor at our center in Kentucky, much to the consternation of the Kentucky Department of Tourism, takes people both to sustainable forestry areas and to an unreclaimed strip mine, to some of the most eroded land in the world due to off-road recreational vehicles, to the Wildcat Mountain Battlefield that is about to be destroyed by developers, and to a U.S. Forest Service so-called shelterwood site, that is actually a clear-cut abomination. Immediately after we initiated our tours, the Forest Service promptly closed the access road to the last site to hinder our activities.

5. I have in one sense experienced passing through some of these stages. I grew up in a pre-EPA, Rachel Carson era with a home life that appreciated the rugged Kentucky countryside and the land that gave us most of our food. Beauty reigned supreme, especially in one small cove away from everyone where I swore I'd fight rather than see it destroyed. Today a limited-access highway runs right through the site. I never fought that battle even though I've performed a quarter of a century of advocacy and environmental demonstration work. In our defeats on numerous occasions I have experienced the powerlessness of public interest work. These have been moments when the environmental degradation of our locality and planet became evident. The need for God's help and prayer to restore the damaged sections of our world is only now striking me. Can I, or rather "we," save our earth?

6. I venture into a feminine critique but volunteer this as a sower of seed for thought. Our experience in the mentioned Earth Healing Program is that women's religious institutions and organizations are so much more responsive to assessments than are men's. Why have we performed about thirty audits of women's religious groups and none of men's, even though we have tried? My hypothesis is that the initiation of genuine earth healing is a feminine ministry just as important as any appropriated by males. As a charter member of Priests for Equality, I do not know where this will lead me, but there may be specific women's spiritual ministries of which earth healing is one. "From the beginning till now the entire creation, as we know, has been groaning in one great act of giving birth" (Rom. 8:22). The recognition of this birth event may be a feminine calling and part of the inherent complementarity of the masculine and feminine for ushering in a new creation.

7. I returned to the only Kentucky town 100 percent below the poverty level in 1983 after a trip to Peru and found the humble little country general store overwhelming in luxury in comparison to what I had just seen. America certainly has luxury, but does it have the good life?

8. A more biocentric approach focuses on the unparalleled growth of human population, mostly in underdeveloped parts of the globe, and projects such frightening scenarios as 14 billion people by the year 2050. While realizing that extrapolations warrant caution, still the health of the planet demands some form of curbing excessive population growth through appropriate means. However, consumption of resources in developed nations remains the single greatest source of global environmental destruction. As the bishops' pastoral letter of 1991, "Renewing the Earth," states, a child born in the United States puts a far heavier burden on the world's resources than one born in a poor developing country. It also points out that we in the first world have only barely begun to curb our consumption.

Part III | IN A DIFFERENT VOICE

May all I say and all I think be in harmony with thee,
God within me, God beyond me, maker of the trees.

—Chinook Psalter

Waking up this morning, I smile,
twenty four brand new hours are before me.
I vow to live fully in each moment
and to look at all beings with eyes of compassion.

—Thich Nhat Hanh

The day of my spiritual awakening
was the day I saw
and knew I saw
God in all things
and all things in God.

—Julian of Norwich

Moving beyond some of the language and imagery found within Judaism and Christianity, the next three authors articulate ecological visions arising from spiritual traditions that are not part of the dominant culture. Writing from their particular vantage points within Buddhist, Native American, and ecofeminist life and praxes, these authors speak in a different voice, challenging the reader and indeed the whole of humanity to reclaim and reaffirm the radical interconnection and interdependence that fundamentally characterizes all of life.

Stephanie Kaza, professor of environmental studies at the University of Vermont, begins this section with a compelling description of the "pattern of domination" that she believes undergirds Western behavior, attitudes, and actions. Thoroughly codified within the current political, social, and economic fabric of society, this oppressive framework justifies a cycle of domination and subordination, from which it is difficult to break free. Environmental degradation and suffering, according to Professor Kaza, are merely consequences of this "logic of domination," and to bring an end to such suffering involves breaking the "gridlock of domination."

In the midst of such suffering, Kaza offers the liberation praxis of Buddhism as one path toward transformation that provides a means for restructuring the fundamental paradigms on which personal, social, and environmental relationships are constituted. Here the Buddhist emphasis on interconnection and interdependence, as exemplified by the Jewel Net of Indra, provides a means of reorienting ourselves in relation to nature and humankind. She sees this view of a truly relational universe as providing human beings an essential starting place for critiquing structural oppression while also making way for a deeper sense of "belonging" and connection.

Through love and the intentional practice of compassion and loving kindness (*karuna* and *metta*), old patterns of domination can be changed and new ways of being in the world can be realized. Kaza concludes by noting that the environmental crisis is a deeply spiritual one, which requires that love become visible through the practice of mindfulness, restraint, and the cultivation of spiritual friendships through which joy and liberation will find their way back into the fabric of human-environmental relations.

In the following chapter, Rick TwoBears, Abnaki elder and Native speaker, writes with the tone of a storyteller, weaving a picture of a universe that is essentially interconnected. His words quickly reveal that the spiritual ecology of which we speak is less a belief system and more a way of being. At the outset he speaks of finding the sacred, not through theories of origins or utilitarian analysis but through experiencing the "miraculous reality and the unfolding aliveness of everything." Such an encounter, he claims, forever changes one's way of seeing.

Once able to see with new eyes, writes TwoBears, one undertakes the spiritual work of letting go of the separate isolated ego in favor of a self that is connected to the whole. Through this process the soul develops, and one becomes increasingly aware of the interconnection of all life. For Two-Bears, it is from these experiences of perceiving sacredness through interconnection and seeking to live accordingly that one begins to learn how to love. It is this love that ultimately provides the greatest sense of wholeness and connection. But he goes on to emphasize that love is a discipline that requires suspending judgment and criticism. Begin simply by loving a plant, just as it is, until with practice you are able to "love all of creation . . . every entity just as it is."

TwoBears closes by describing the role of the drum within Native life. A gift from women, the birthers and nurturers, the drum serves as bible, sacred symbol, living witness, and rhythmic partner and plays a vital role in revealing the interconnective web, the true community of which all are a part.

Catherine Keller, professor of theology at Drew Theological School in Madison, New Jersey, offers a third "different voice." Keller reminds the reader from the outset that feminism today, as she has come to know it, clearly includes an ever broadening array of justice issues. At its core, ecofeminism is about interconnection and is a call to rediscover the radical interdependence that is the essence of life.

Professor Keller, using the metaphor of the compost heap, explores the ways in which patriarchal culture has denied the presence and power of decay and death. Within such a culture the composting processes represent chaos and are a threat to the desire for order and control. Everything that is of death and impermanence, everything that stands as a reminder of the passing materiality of this world, is devalued, cast off, and abandoned in favor of an immutable, nonearthbound eternity. Yet, asserts Keller, it is there on the compost heap that transformation occurs, that waste, decay, and death become the seedbed for a future alive with newness and possibility. She invites the reader and all who care for life to be more attentive to waste and subsequently to finitude, undertaking the holy work of sifting and sorting through the refuse of history and our inherited social order with the hope of determining what can actually be recycled and what must be eliminated.

Expanding on the metaphor of the compost heap, Keller, who sees negative environmental consequences in using language for deity that is anthropocentric and transcendent, speaks of "God" as recycler and recycled, that "sacred matrix of life" that is not beyond or outside life but is both the transforming and transformed presence connecting all things. "Such a divine process of recycling," she writes, "suggests . . . that to waste . . . our wastes, to disparage material life and therefore to destroy it, is to go against the grain of the universe." To join in the great recycling process is to engage in holy work that will necessarily reconnect us to a quality of life and relationship that will deepen our experience of living and draw us into the sacred rhythms of the changer and the changed.

8 | STEPHANIE KAZA

The Gridlock of Domination

A Buddhist Response to Environmental Suffering

Sakyamuni, the Buddha-to-be, was born around 560 B.C.E. in the wooded garden of Lumbini near Kapilavastu. There, below the towering peaks of the Himalayas, deep in the watershed of the River Ganga, young Siddhartha began his princely life. According to the legend, Prince Siddhartha's father protected him from exposure to any sign of sorrow or suffering.[1] His first encounters with old age, sickness, and death moved him to meditate on the pervasive truth of suffering. Leaving his home and family, he set off on a spiritual quest for understanding. The canonical account describes Gautama's study under many teachers, with extensive austerity practiced even to the point of starvation. Realizing the limits of deprivation, he sat by a sacred Bodhi tree to gain strength, accepting food offerings from a woman named Sujata.

As the story goes, he stayed by the Bodhi tree for seven days, resolved not to arise until he gained deep understanding. During the seventh night he was tormented by every possible distraction of the mind. Mara, the voice of delusion, challenged him ferociously, asking what right he had to sit by the tree, seeking the truth. To counter the force of ignorance, he touched his right hand to the earth, calling for witness. At this moment, with earth and tree in witness, the Buddha realized enlightenment, seeing clearly the interdependent nature of all reality.

The Buddha recognized that all life was characterized by impermanence—the constant arising and passing away of phenomena. He understood that as long as people were caught in the endless round of birth, sickness, old age, and death, there could be no end to suffering. The Buddha today might define the "good life" as a life free of suffering. But he realized that suffering can not be extinguished, that the only path to freedom lies in the midst of suffering. It is right here, he suggested, that one experiences the deep truth of the interconnectedness of all phenomena.

The Buddhist path to the good life can be seen as a practical method for personal and social transformation. This path offers a liberation praxis based on the profound act of waking up to the nature of reality. Through mindfulness, meditation, and morality practices, one cultivates the mind of wisdom and compassion. Through direct knowing, one experiences the truth of interdependence. These Buddhist practices are particularly useful in waking up to the state of widespread environmental suffering and deteriorating human–nature relations.

The early Pali texts recount the Buddha's concern for the environment in both parable and precept.[2] In the Jataka Tales, stories of the Buddha's past lives, the Buddha shows great compassion as a suffering monkey king or patient buffalo.[3] The precepts or guidelines for monastic life include care with water use, simplicity in housing, respect for animals, and special attention to trees. In following the precepts, he preached, one must take care of one's actions in three arenas: body, speech, and mind. Restraint from harming another meant not only physical and verbal restraint but also mental restraint. The Buddha felt that even having the thought of harming another being carried karmic consequence.

From a Buddhist perspective, every action toward a plant, animal, or any being consists of three parts: the intention of the act, the thought of the act, and the actual doing of the act. The fullness of this conception is critical to an accurate understanding of "environment" in Buddhist philosophy. The original Japanese word for nature, *shizen*, meant "what is so of itself," the realm of spontaneous becoming.[4] The "suchness," for example, of a flower is the unfolding dynamic of seed to bloom, the ripening of pollen and fruit, the process of decay—all of which contain the long evolutionary history of the species, the cumulative events of the season and place, and the plant's particular spatial and ecological relationships. The Japanese experience of nature includes the interpenetrating aspect of mind, both in the perception of the experience and in how it shapes one's actions toward the environment. To approach a conversation about the environment in Buddhist terms, one must recognize in a dynamic way the profound influence of human thought on perception and action.

In this chapter I will focus on one paradigm of thought and action that has many ramifications for the environment. That is the pattern of domination—domination of gender, race, class, and religion and domination of nature. I will draw on work by feminist scholars and environmental justice advocates as well as my study and practice of Zen Buddhism. I believe the pattern of domination manifested in both attitudes and actions is a central grid underlying much of human behavior toward the natural world. The objectifying perception of trees as board feet and rabbits as cosmetic test animals holds much in common with the mind that justifies hazardous waste sites in racial ghettos. The act of clear-cutting large expanses of forest for paper products has much in common with the act of clearing an inner city neighborhood to build a new shopping mall. Dominating attitudes and actions transcend class and culture, affecting environment and social justice issues across the continent—in urban neighborhoods, rural farmlands, Indian reservations, and middle-class shopping malls. Any meaningful discussion of the environment today must include social justice issues, for they are inextricably intertwined with environmental degradation. The thread of domination runs throughout the current weave of habitat destruction, cultural invasion, economic competition, and human rights violations.

As an American Buddhist, I want to test this religious heritage for spiritual relevance today. I want to know if it can make a difference in the terrible specter of destruction on the landscape. Certainly, we need as much help as possible in turning the destructive tide of species loss, habitat destruction, and runaway consumption of resources. I believe that Buddhist practice and philosophy may be particularly helpful in untangling the widespread pattern of domination. The powerful liberation praxis of Buddhism may provide the radical transformation necessary to alter the downward spiraling course of environmental history.

Patterns of Domination

Let me begin with some examples to illustrate the common features of the dominating mind. These stories are painful to hear, as is often true in facing environmental realities. In each case, one or more parties are being systematically harmed by another who holds more power. The players vary, but the patterns of behavior are similar. Each story is a complex web of economic, political, and psychological power dynamics working together to cause human and environmental suffering.

The first example is from the rural back country of northeastern North Carolina. Purdue Chicken has located a number of new chicken factories in these poor areas, where permits are cheap and labor laws lax. The animals are raised in factory-farm conditions, sometimes up to fifty thousand in a building, five hens to a cage, barely able to turn around.[5] Their wings and beaks are clipped to reduce damage. Most of the processing is done by uneducated, unskilled African-American women to support their families. The women stand in blood and ice water as the slaughtered chickens come down the line. They eviscerate, debone, and pack chickens at a rate of 90 to 120 birds per minute. According to the Center for Women's Economic Alternatives, 79 percent of the work force suffers from some form of degenerative health problem due to the pressure of repetitive motion and poor working conditions.[6] Many develop arthritis in their hands by the age of twenty-seven, severely limiting their chances for other work later on.

The second example comes from just north of Oakland, California, on the edge of San Francisco Bay in the city of Richmond—population 80,000, over half African Americans and 10 percent Latinos. Most of the African Americans live near the Bay Area's concentrated petrochemical corridor of 350 facilities handling hazardous waste. Some of the largest toxics producers are the Chevron oil refinery, the Chevron Ortho pesticide plant, Witco Chemical, and Airco Industrial Gases. Chevron Ortho alone produces 40 percent of the hazardous waste in Richmond, most of which is incinerated at the factory (approximately 75,000 tons per year). Toxic emissions regularly pollute the air and water in the area, causing chronic illness, especially among children. Citizen organizers of the West County Toxics Coalition are negotiating a six-point plan with Chevron, proposing a 1 percent clean-up fund and a twenty-four-hour public health clinic for those affected.[7]

A third example lies in the two-thousand-mile border area along the Rio Grande River between the United States and Mexico, where over 1,900 *maquiladora* assembly plants employ a half million Mexicans. The plants are operated by American, Japanese, and other foreign corporations, who gain a profit advantage by using cheap labor. Mexican workers have flocked to the border for these low-paying jobs, placing severe strains on sewage and water systems. Worker and resident health is compromised by poor air quality and health care. Environmental regulations across the border are weaker and less well enforced, assuring reduced business costs for major corporations.[8]

Fourth, in South Dakota, local authorities still interfere with the Native American Sun Dance worship practice in certain places, a holdover from missionary edicts. The Sun Dance ritual celebrates the place of the sun in sustaining the lives of the people, as well as of the plants and animals that support them. It is a key rite for the Crow, Shoshone, and Sioux people, reaffirming the sacred trust between people and the earth, maintaining the foundation of an earth-based spirituality. Despite passage of the Native American Religious Freedom Act in 1978, many tribes are only beginning to reestablish the community forms of their earth-based spiritual traditions.[9] For more than a century these so-called pagan celebrations were barred by Christian preachers who questioned the existence of souls in plants and animals.

The fifth story comes from my homeland, the Pacific Northwest. The Columbia River basin was once covered with thick stands of Douglas fir, red cedar, western hemlock, and ponderosa pine. Intensive logging on public lands began during my childhood. Now one can fly from southern Oregon to Vancouver, British Columbia, in a low-flying plane and never be out of sight of bald, scarred clear-cut holes in the forest landscape.[10] Under congressional and timber industry pressures to "get out the cut," the once magnificent northwest conifer belt has been reduced to fragments. Ninety percent of the original old-growth Douglas fir forest has been turned into houses, pulpboard, and paper. What remains is now subject to intense negotiations between the White House, international timber megacorporations, and environmental groups. President Clinton's Option 9 plan emphasizes watershed management to reduce erosion, but it does not protect the remaining old growth from cutting. A large old-growth tree is now worth up to five thousand dollars; an old-growth redwood may fetch up to ten thousand dollars—the incentives for harvest override most other considerations.

What do these five examples hold in common? I submit that they are all shaped by the same oppressive conceptual framework, played out by economic and social forces across the continent and globe—the framework that justifies and explains the pattern of domination. In every case, one point of view or class of people prevails over another: corporate business over labor, profit margin over worker health, colonial culture over indigenous culture, managed lands over wild lands. In every case, the suffering of trees, animals, and people is rationalized as necessary by the dominating party. To me, it is quite clear that this oppressive conceptual framework is a significant barrier to sustainable and harmonious existence with the environment.

The Logic of Domination

Let us look now at how this pattern of domination works through attitude, action, thought, and behavior. It is helpful to distinguish between inter-

personal and structural, or systemic, patterns. One might associate domination with the physical act of rape or emotional displays of anger that
occur on the individual level, but that type of domination is inadequate to
explain the wide-scale acts of destruction to the environment across continent and nation-state boundaries. Instead we must examine structural
patterns of domination codified in government institutions, private enterprise, and global economic exchange. These are most identifiable in human
labor practices, corporate acquisition of natural resources, and free market
justification of ecological destruction.

Consistent throughout systemic patterns is the logic of domination.
Ecofeminist philosopher Karen Warren lays out the steps in this logic as she
describes the parallels between domination of women and domination of
nature.[11] All societies construct and reinforce specific perceptual lenses
through which its members recognize their relationship to the larger group.
A conceptual framework consists of basic beliefs, values, attitudes, and
assumptions thought to be held in common. For example, the framework
called democracy includes the value of equal opportunity, shared electorate
decision making, freedom of speech and assembly, and equal rights for
minorities. An *oppressive* conceptual framework is one in which the beliefs
and values "explain, justify, and maintain relationships of domination and
subordination."[12] The framework called racism, for example, assumes that
one race is superior to another, usually white to color, and this belief
justifies lower wages, police brutality, and sexual harassment of people of
color.

As described by Warren, an oppressive framework can be recognized by
three significant patterns of thinking. The first is the type of thinking that
places higher value or status on that which is perceived to be "above"
something "below" it. Animals are commonly thought to be more advanced than plants, mammals more intelligent than reptiles, European
culture more civilized than Native American. The second is the use of
mutually exclusive pairs to represent value. For example, white is seen as
opposite to black, reason as opposite to emotion, male to female, mind to
body. In each case, members of the pair are perceived as noninclusive: each
element does not contain the other. Usually, one element of the pair is
given more social or moral value than the other. In contrast to this traditional Western use of dualisms, Eastern philosophical traditions perceive
such pairs as complementary and part of a mutual whole. A strong visual
symbol of this is the yin-yang circle, where the black half contains a seed of
white; the white half, a seed of black.

The third feature and key to the logic of domination is that one half of a dualism is assigned moral superiority by the dominant social group. The group then agrees (consciously or unconsciously) that superiority justifies domination. Without this logic, differences between genders, ethnic groups, ways of knowing, or landscapes would be perceived simply as that—differences. Yet we have general Western agreement that humans are morally superior to animals and plants and thus are justified in using them for their own ends. Or that black and Hispanic lower classes are somehow morally inferior to white middle and upper classes, thereby justifying the placement of toxic waste sites in their neighborhoods.[13] Or that tree plantations are superior to complex old-growth forests because of their usefulness to human beings. Or that earth-based religious traditions are inferior to more "civilized" Western traditions. In every example of environmental disaster or degradation, one can see the logic of domination functioning behind the decision-making process. I submit that one of the most profound contributions of modern religious traditions could be the exposure and abolition of this logic of domination in all its insidious and environmentally destructive forms.

A Buddhist Response

The Buddhist spiritual path is based in the practice of liberation in the midst of suffering. The word *Buddha* means "awakened one." The path to liberation is not a way to escape the suffering of the world; rather it is a way *through* the suffering to inner spiritual freedom. In the case of domination the suffering is experienced by both dominator and dominated. It is experienced by individual people, plants, animals, and places, and it is experienced collectively by social groups and ecological systems. These forms of suffering are different but intimately related. For example, the suffering of overconsumption and addictions to excess in the North depend on the suffering of poverty and cash crop economies in the South. Buddhist liberation comes through understanding the nature of domination and being moved by compassion to alleviate the suffering of others, whether personal or collective.

The goal of freedom in Buddhism liberation is significantly different from the goal of freedom in the Western political worldview. For example, let's look at freedom in the context of "free enterprise." Free enterprise was a key concept in the hotly debated NAFTA trade negotiations, which some say favored economic gain over environmental health. The word *free,* as used in Western discourse, carries the assumption that anyone should be able to do whatever s/he needs to achieve economic success. The underlying premise is that everything is possible for everybody, that there is a fundamental universal equality underlying all human existence, that given the right opportunities, a person can rise to any level of achievement.[14] The implication is that a person can be "self-realized" or liberated *independent* of context and relationship. This concept of freedom often is interpreted as freedom *from* relationship.

The Buddhist view, in contrast, starts from the assumption of relationship, defining freedom within this context. Each person exists and acts in a web of relationships that have developed through actual historical events. Each person's suffering is seen as a manifestation of specific events, attitudes, and contexts. This is often described in terms of *karma*—the law of cause and effect that holds one accountable for the impacts of one's actions in the world. Rather than beginning with an idealized version of human existence as free, Buddhism begins with the reality of each individual's specific suffering. Buddhist philosophy emphasizes that *all* points of view are conditioned by individual experience; the challenge of liberation is to investigate fully the nature of this conditioning.

The Buddha's method for liberation is described in the Four Noble Truths. The first truth establishes the existence of suffering due to the impermanent nature of all things. Birth, sickness, old age, and death come to mountains, mountain lions, mountain walkers, and mountain bikes alike—all are impermanent. By opening to suffering, one gains a direct and moving experience of the nature of existence. A Buddhist sensitized to Warren's insight into the logic of domination will recognize the pervasive patterns of suffering caused by domination. Clear-cut landscapes, oil-covered cormorants, polluted beaches—experiencing environmental death and sickness as physical manifestations of the dominating mind can be a first step toward awakening.

The second truth is that the cause of suffering is ignorance, which gives rise to greed, hate, fear, and other painful mental and emotional states. This means physical and ecological ignorance as well as ignorance of the dualistic, objectifying, dominating habits of the conditioned mind. It is these states of mind that cause people to dominate plants and animals as well as other people. For example, harmless snakes are killed out of ignorant fear of poisonous snakes; native tropical rain forests are cut and replaced by exotic eucalyptus plantations for the greed of economic profit. In the first case, individual humans dominate over snakes, determining their fate. In the second case, corporate business preference for fast-growing timber dominates over the complex biodiversity and local economics supported by the forest. Through Buddhist spiritual practice one can examine the consequences of environmental ignorance and see in detail the causes and conditions that perpetuate damage to the earth.

The third truth is that there can be an end to suffering, that one can find liberation from the traps of the conditioned mind. In an ecological sense, this means liberation from the dominating habits of human superiority, of the privilege of waste, of unconscious racial, ethnic, and gender bias. On a landscape scale, liberation must also be from local and national delusions of environmental autonomy. In fact, the movement of air and water transcends political boundaries and must be dealt with as a flowing medium for all human activity.

The fourth truth contains the Buddha's specific prescription for freedom. The path of liberation is based fundamentally on two things: wisdom and compassion. These are manifest in the world through the Eight-Fold Path, which includes practice arenas such as of right livelihood and right speech. By working mindfully to understand the pervasive conditioning of domination, one finds myriad opportunities for waking up and liberating the self and others. This may be through environmental justice action, forest protection, hunger relief, or any number of related activities.

Insight Wisdom or Understanding

Buddhist insight wisdom is not understanding that can be attained by reading a book. It is less a theory to be learned in a conventional sense but rather a truth to be experienced. This fundamental teaching is what the Buddha discovered sitting under the Bodhi tree: the law of dependent co-arising, or *paticca samuppada*. This is the great truth that all events and beings are interdependent and interrelated. The universe is described as a mutually causal web of relationship, each action affecting many others in turn. An image for this cosmology from the Mahayana Buddhist tradition is the Jewel Net of Indra. This multidimensional net stretches through all space and time, connecting an infinite number of jewels in the universe. Each jewel is infinitely multifaceted and reflects every other jewel in the net. There is nothing outside the net and nothing that does not reverberate its presence throughout the web of relationships.[15]

From an ecological perspective, this law is obvious. In example after example, it is painfully apparent that ecological systems are connected through water, air, and soil pathways. Chemical pesticides on agricultural lands run off into adjacent wetlands, affecting wildlife reproduction; sulfur emissions from industrial centers acidify rain falling on spruce forests several watersheds away; drifting shipboard plastics strangle gulls and seals on island breeding grounds.

From a Buddhist perspective, interdependence also includes the role of human thought and conditioning in the mind. This stands in contrast to a scientific interpretation of interdependence, primarily of the "external" world. Buddhists see human thought as both shaper of and shaped by the ecological crisis. For example, schools of business and natural resources train professionals with the maxims that profit making is central, that cheap labor is preferable, that game management requires ecological manipulation (adding fish stocks, culling deer herds, regulating predator populations). These ideas determine actions, and the actions in turn reinforce the mental conditioning. I would suggest that domination is one of the most prevalent and destructive patterns of mind in the twentieth-century race for resources.

There are two interrelated aspects of the law of dependent co-arising; each carries important implications for understanding patterns of domination. The first is that all perceptions, phenomena, and worldviews are *impermanent*. Nothing is absolute and unchanging; nothing endures. What all things hold in common is that they arise and they pass away—whether one speaks of insects, ice fields, cultures, or galaxies. Being nonabsolute, all phenomena are relative and conditioned by the context in which they have arisen. There is no such thing as a permanent ice field, for example; it is continually changing shape by melting and moving, subject to long-term weather shifts and the shape of local topography. Cultures too are conditioned and shaped, by dietary habits, soil types, religious ceremonies, wars, and other passing phenomena. Likewise, patterns of domination are also impermanent and conditioned, held in place only by the mutually causal web that sustains them. Both actions and attitudes of domination can be taken apart, systematically disarming the web of agreements that holds them in place.

The second aspect of this law is that all phenomena are *co-dependently produced*; nothing arises in the universe independently. Even one's own existence is not a separate, autonomous existence. This is a radical departure from the Western view of the self as autonomous and motivated by will and self-interest. Buddhism identifies the primary delusion of mind as false reification of the self as an enduring, independent existence. Buddhist practices emphasize breaking through the myriad expressions of this delusion of separate self to fully realize the nature of reality as interdependent. In this experiencing of truth, one's most deeply conditioned assumptions about the conventional dichotomies between self and world are radically transformed. Here is the moment of liberation, the opportunity for hope that there can be another way of relating to the environment.[16]

How does the law of dependent co-arising apply to patterns of domination? Every instance of domination—of forests, indigenous peoples, or urban poor neighborhoods—reflects patterns of social and personal conditioning. These patterns are made up of power relations, historical contracts, economic motives, human fear of risk and difference, and threats to survival. Each of these can be investigated in depth for insight understanding. For example, looking deeply into the logging controversy of the Pacific Northwest, one sees not simply loggers and owls but also outdated forest management practices; congressional pressure for unsustainable cuts; loss of salmon runs from erosion-silted streams; escalating increase in paper use for computers, copiers, and fax machines; a national economy built on material progress. The web is further complicated by ethical questioning from within agency ranks and increased citizen concern for the fragmented landscape, as well as clashes in assumptions about the nature of "nature." Reviewing Warren's features of domination, we can see hierarchical thinking in class differences between loggers, environmentalists, and politicians. Charged dualisms include loggers versus owls, forests versus jobs, human needs versus animal and plant needs. The traditional logic of domination attributes greater moral weight to people over trees; environmentalists and concerned citizens challenge this. The conflict is riddled throughout with patterns of dominating thought and action.

Thus, from a Buddhist perspective, domination is not some abstract force that one can eradicate or outlaw by will; rather it is the sum of all the mutually reinforcing interactions that are oppressive in any given situation. Each of these can be an arena for awakening by seeing into the patterns of conditioning. Right within the suffering of domination lies the possibility for liberation.

Some common habits of mind that reflect domination of the natural world are stereotyping and projection. Each of these reinforces perceptions of beings as objects rather than as members of a web of relationships. With *stereotyping*, people oversimplify and lump a few characteristics of an animal or ecosystem into a generic representation. For example, whales are commonly seen as playful, altruistic, intelligent, large, and gentle—each characteristic fitting one species or another but not existing anywhere in this combination in a real whale. Thus, the stereotype of whales dominates over the specific reality of individual species and organisms. In *projection*, the mind projects internalized values of good and bad onto favored and unfavored elements of the environment. "Good" land is land that can be farmed or developed; "bad" land is what is too steep, dry, or impenetrable to be subdued.[17] Human ideas of land values dominate over ecosystem values; the taming force of cultivation comes to dominate over the vitality of wilderness.

Let us look at one example that lies at the heart of the environmental crisis: runaway increase in human population. Increasing pressure on natural resources is often cast as the fault of exploding populations in countries of the South. One suggested solution is birth control for women in those countries. By implication it is those women who are responsible for the environmental predicament. This perspective has prevailed in most international development arenas for the past twenty years; it reflects a northern industrial point of view. At the 1991 World Women's Congress in Florida and subsequently at the Earth Summit in Brazil and other international conferences, women of the South expressed outrage at the dominating nature of the northern perspective.[18] They point to the enormous rate of resource consumption by countries of the North, laying the blame for environmental abuse at the feet of a wasteful consumer society. They resent the neocolonial wave of North-directed solutions to population control, demanding instead self-determination and support for local reproductive health clinics. The northern perspective assumes that a solution must be imposed on the less well-off South, with little recognition of the North's disproportionate use of resources. The women of the South are resisting the violence of colonial domination, first applied to their natural resources, then to their labor, and now seen as a challenge to their own bodies.[19] The North, meanwhile, is caught in the domination of excess consumption, corporate capitalism, and the combined forces of advertising and the media.

A Buddhist approach to population and consumption would seek liberation for both the overpopulated Southerners and the overconsuming Northerners. Indian and African women, for example, point out that reproductive rates drop when women gain economic and educational freedom. Therefore, the simplest and most empowering route to population control is education and local self-reliance initiatives. These methods bring control to women directly, taking it out of the hands of well-meaning but patronizing dominators. Environmental activists urge the North to evaluate its own levels of resource consumption and hold accountable those with runaway power, such as large multinational corporations. Some have suggested that twelve-step programs for substance abuse be extended to investigate addictions to petroleum, paper products, and recreational shopping.

Applying the Four Noble Truths as a framework for analysis, one can pose four questions: (1) What is the problem or suffering? (2) What are the causes of the suffering? (3) What would put an end to the suffering? (4) What is the path to realize this goal? The problem can be described as interlinking dominations, with patterns in the North and South each exacerbating the other. The causes are historical, cultural, topographical, and interpenetrating over time. I am suggesting that taking a stand against domination in all its various aspects would dramatically shift national and international choices for reducing the environmental impact of increasing population and consumption. This path is not yet charted, though activists on both sides have suggested first steps.

The spiritual strength of this approach lies in its emphasis on recognizing and addressing suffering. In this way, those engaged with the issues develop heartfelt relationships with those who are suffering—plants, animals, and places as well as people and cultures. This analysis provides an alternative to technological or politically driven approaches that emphasize material-based solutions and avoid an emotional response to suffering. Applying the law of dependent co-arising, the population/consumption problem takes on new dimensions and also new possibilities for solutions. This is practicing the heart of Buddhism with a small *b*, as Thai activist Sulak Sivaraksa says.[20] By this he means practicing awareness and responding directly to suffering, not necessarily from within the self-constructed identities of organized religion.

The Path of Compassion

Facing tremendous suffering experienced by diverse cultures and species can be devastating without a fundamental grounding in love and compassion. The two pillars of Buddhist practice, wisdom and compassion, complement and strengthen each other. One can approach domination through inquiry and be moved by the complexity and pervasiveness of suffering. Or one may first be touched with compassion for the suffering— a bloody harp seal, a child's swollen belly, a ravaged forest—and then seek understanding.

The Buddha encouraged two kinds of love: *karuna,* or compassion, and *metta,* or loving kindness. *Karuna* arises as the natural and spontaneous response of the heart to the suffering of others, often felt as the desire to help alleviate their pain.[21] The capacity for compassion grows through opening the heart to circumstances filled with difficulty. One realizes with compassion that each party carries not only its own suffering but *all* aspects of what is involved. A Buddhist approach to domination of people and the environment evokes compassion for the diverse and specific suffering of *all* participants, seeking solutions that reduce suffering in as many arenas as possible. This would include compassionate actions and also encouraging compassionate attitudes toward the others involved. A Buddhist version of an environmental impact report, for example, might include a long list of those already suffering and those who might suffer from a proposed dam or development. A Buddhist conflict resolution might begin with a round of listening to each party's suffering.

Applying compassion to the arena of domination, one must include examination of oneself for all aspects of the dominating relationship. One quickly uncovers the mind's tendency toward dualistic thinking, the tendency to polarize a situation into dominated and dominator roles, to find a party to support and a party to blame. One also finds the urge to sink into psychic numbing, denial, fear, rage, or despair—all of which can block effective action.[22] But all of these are too simplistic. Each promulgates the delusion of a separate self-identity: the good person, the bad person. One sees that *no one* is free of the grid of domination. For example, I can name myself as dominated because I am a woman, subject to job discrimination and sexual harassment. Yet as a white privileged person of the North I am perceived as dominator by those in the South. As a tree lover, writing about conversations with trees, I sometimes feel marginalized by mainstream values that objectify trees.[23] Yet I am a dominator of trees in my own endless use of paper. One sees there is no absolute position free of suffering. Each person in the interdependent web carries the implication of all actions. In the midst of this, one cultivates intention to act for change, eyes fully open to the infinite dimensions of environmental suffering caused by dominating mind.

The practice of *metta*, or loving kindness, is more of a proactive generation of positive commitment to the well-being of others, offering love on behalf of oneself, one's friends and family, even one's enemies, and all beings. According to the Theravadin tradition, which emphasizes *metta*, to make such statements on a regular basis develops the habit of thinking positively of others, whether plants, animals, or people.[24] This holds the possibility of being a strong antidote to *anthropocentrism*, the habit of centering one's thoughts and actions around human needs rather than the wider needs of all beings. Loving kindness can be an actual force of renewal, challenging the pervasive paradigm of domination and planting seeds of joy that may support a life of ecological sanity and sustainability.[25]

A Buddhist approach to environmental conflict resolution is fundamentally based on *metta* and *karuna*. When people are motivated by a genuine desire for harmonious relations, they bring a creative openness and refreshing willingness to the problem at hand. One person's capacity to refrain from dominating mind can change the dynamics of a whole group. In combination with the compassionate force of Buddhist analysis, the practice of *metta* can generate forward movement in the midst of great crisis. As His Holiness the Dalai Lama points out, "When we talk about preservation of the environment, it is related to many other things. Ultimately, the decision must come from the human heart. The key point is to have a genuine sense of universal responsibility, based on love and compassion, and clear awareness."[26]

Everyday Practice

If patterns of domination are conditioned and impermanent, they can be taken apart. With the tools of wisdom and compassion, spiritual practice can focus on the deconstruction of these environmentally devastating patterns of thought and behavior.

It is increasingly clear to me, as it is to many, that the environmental crisis for humans is at heart a spiritual crisis. Great strength of spirit is needed to respond to the overwhelming dimensions of environmental destruction. All religious traditions have the capacity to awaken and inspire the human spirit. I believe that three principles of Buddhist everyday practice and philosophy can make a major contribution to this awakening.

The first is the practice of mindfulness: becoming aware of what one is actually doing. Under the pressures of modern life, one of the most difficult tasks is to simply "be present." In conversation and mental activity, the practice of mindfulness brings attention to one's words and thoughts as they affect others. In environmental conflicts one sees the socially conditioned tendency of the mind to create enemies, to lapse into domination and control. Paying attention, one can practice restraint from reactivity, stereotyping, and projection. Slowing down, one can include attention to the larger interdependent web. This means embracing the earth one walks on, the air one breathes, the food one eats.[27] By practicing mindfulness the spiritual practitioner increases the odds for waking up in the midst of environmental suffering.

The second practice is restraint. By not perpetrating patterns of domination, one reduces their momentum and power. This may be practiced in both mental and material realms. Stopping the habit of blame and stopping the habit of imperialism both require mental restraint. In the material realm, Buddhism has long advocated a simple life, dependent on few physical necessities. The choice to reduce consumption of material goods can be a spiritual practice based in the desire for greater awareness. Reducing time spent in dealing with objects (cars, videos, television, toys) can allow time for actively cultivating intimacy with rivers and mountains, moon and stars.

A third practice is the cultivation of spiritual friendships. Developing relationships free of domination is a spiritual task that can only benefit human–environment relations. One can seek spiritual friendships with trees, animals, and places, as well as with people. One of the Buddha's students once remarked that it seemed to him that having good friends was half of the holy life. "Not so," the Buddha replied. "Having good friends is the whole of spiritual life." As spiritual friends in a complex, ecological world, we can help each other wake up to the splendid, if troubled, multidimensional reality of the environment. I believe it is our spiritual task to help each other bear the generational responsibility for turning the tide of ecological destruction. Practicing mindfulness and restraint in the company of spiritual friends, we can maintain our compassion and wisdom, finding the courage to act boldly. Acting from the truth of interdependence, we can plant the seeds of joy that will nurture sustainable spiritual relations with all beings far into the future.

Notes

1. Richard H. Robinson and Willard J. Johnson, *The Buddhist Religion* (Belmont, Calif.: Wadsworth, 1982), 7–14.

2. See Chatsumarn Kabilisingh's survey of early texts for Wildlife Fund Thailand: "Early Buddhist Views of Nature" in *Dharma Gaia*, ed. Alan Hunt-Badiner, (Berkeley, Calif.: Parallax Press, 1990), 8–13.

3. *Twenty Jataka Tales*, retold by Noor Inayat Khan (New York: Inner Traditions International, 1985).

4. Hubertus Tellenbach and Bin Kimura, "The Japanese Concept of Nature," in *Nature in Asian Traditions of Thought*, ed. J. Baird Callicott and Roger T. Ames (Albany: State University of New York Press, 1989), 153–81.

5. See Peter Singer's *Animal Liberation* (New York: Avon, 1990), 98–119, for detailed documentation of chicken-raising conditions.

6. Stephanie Kaza, "Another Reason for Being a Vegetarian," *Turning Wheel*, Spring 1993, 8.

7. Robert D. Bullard, *Confronting Environmental Racism: Voices from the Grassroots* (Boston: South End Press, 1993), 29, 35.

8. Ibid., 19.

9. The Sioux Sun Dance ceremony is described in detail in Ed McGaa, *Mother Earth Spirituality* (San Francisco: Harper and Row, 1990), 85–96; the trauma of cultural conflict with Christian missionaries is described in Sam D. Gill, *Native American Traditions* (Belmont, Calif.: Wadsworth, 1983), 169–72.

10. See the unmistakable photographic evidence in Bill Devall, ed., *Clearcut: The Tragedy of Industrial Forestry* (San Francisco: Sierra Club and Earth Island Press, 1993).

11. Karen Warren, "The Power and Promise of Ecological Feminism, *Journal of Environmental Ethics*, 1990): 12:125–46.

12. Ibid., 127–28. The three features that follow are described on p. 128.

13. See the Cerrell Associates report on the California Waste Management Board in Bullard, *Confronting Environmental Racism*, 18.

14. Gerard Fourez, *Liberation Ethics* (Philadelphia: Temple University Press, 1982), 18–21.

15. Francis H. Cook, "The Jewel Net of Indra," in *Nature in Asian Traditions of Thought* (Albany: State University of New York Press, 1989), 213–30.

16. Robert Aiken, *Taking the Path of Zen* (San Francisco: North Point Press, 1987).

17. Gary Snyder, "Good, Wild, and Sacred," in *The Practice of the Wild* (San Francisco: North Point Press, 1990), 78–96.

18. Gita Sen, "Women, Poverty, and Population Issues for the Concerned Environmentalist," in *Feminist Perspectives on Sustainable Development*, ed. Wendy Harcourt (London: Zed Books, 1994), 215–25.

19. Vandana Shiva, "The Seed and the Earth: Technology and the Colonization of Regeneration," testimony before the World Women's Congress for a Healthy Planet, 8–12 November, Miami.

20. Sulak Sivaraksa, *Seeds of Peace* (Berkeley, Calif.: Parallax Press, 1992), 62–72.

21. See, for example, "Stopping the War," in Jack Kornfield, *A Path with Heart* (New York: Bantam, 1993), 22–30.

22. Joanna Macy, *Despair and Personal Power in the Nuclear Age* (Philadelphia: New Society Publishers, 1983).

23. Stephanie Kaza, *The Attentive Heart: Conversations with Trees* (New York: Ballantine Books, 1993).

24. Joseph Goldstein and Jack Kornfield, *Seeking the Heart of Wisdom* (Boston: Shambala, 1987), 96.

25. Thich Nhat Hanh, *Plum Village Chanting Book* (Berkeley, Calif.: Parallax Press, 1990). The phrase "planting seeds of joy" refers to Nhat Hanh's most recent revision of his Fourteen Precepts of the Order of Interbeing.

26. Dalai Lama, "Ecology and the Human Heart" in *My Tibet*, ed. Galen Rowell (Berkeley, Calif.: University of California Press, 1990), 79–80.

27. Thich Nhat Hanh, *Being Peace* (Berkeley, Calif.: Parallax Press, 1987), 105–15.

Seeing the Whole

I am a member of the Abnaki tribe of Native Americans, a tribe that lived within and roamed the woodlands of what is now called New England. I am not, however, just a descendant but also a practitioner of their way, their spirituality, their way with the earth. When I try to communicate this form of spiritual ecology to non-Indians, I like them to see the whole, the interdependence and interconnectedness of all of life. But then, of course, the question arises: how can I best accomplish that?

As you will soon see, my approach to this topic is not a straight line but, like the universe itself, always circular. It may seem that I am taking my canoe first this way and then that way, and I will sometimes even stop to talk about the loon over there, but eventually I'll return to the canoe so that the interconnectedness of the earth will become apparent. You see, nothing is irrelevant, and the best way to get to the central thread that ties everything together is to follow it wherever it leads.

All of Life Is Interconnected

My people have a story about Little Hawk. Little Hawk was a hunter who, out of concern for his people, went out and gathered, hunted, and fished for far more than he or his tribe could consume. He brought it all back to the village. Then he was asked by the elders, "Why did you do this? Why did you take more from the earth than we need?" Little Hawk replied that he just wanted to store up enough food so that his people would not have to hunt every day.

From one point of view, of course, Little Hawk was perfectly right. It is right and proper to think ahead and to store food for the future. On the other hand, Little Hawk forgot that the "extra" grain and fish and animals he killed to accomplish that were simply destroyed. So the elders pointed this out to him and asked him if he had thought about food for their children, not only this generation of children but seven generations down the line. There is a reason for all things, and Little Hawk in his effort to help his people actually destroyed some of that connectedness.

That is why, for many Native peoples, the hawk has become the symbol of conservation in its deepest sense, in the sense of becoming a guardian to watch over and protect the interconnected flow of nature, to keep that ecosystem in harmony and balance. There is a natural order and connection between all things in nature, an order that is implicit to that nature and not just an imposition of man's conception of order and design. Everything, you see, is interconnected and mutually dependent, now and forever. But how do we come to perceive such an interconnection, and how does it affect how we live and behave?

To See the Interconnection Is to See the Sacred and to Perceive Life Differently

Some time ago a friend told me that his daughter's school had recently invited a Native American who told the students that stones—and in fact everything in nature—are sacred because they "live." I asked him what he thought about that, and he replied, "Yes, of course, they are sacred." Both the students in that school and my friend felt like that because they had actually come to experience nature that way. What I am trying to communicate here is that the awareness of the sacred is an *experience*, not a hypothesis or some sort of demonstrated conclusion. But what is that experience of the sacredness of everything like?

It is a way of seeing in and through and beyond each individual thing to perceive the relatedness and interconnection of each thing to everything else. Finding the sacred is not to have a theory about the origins of things or to apprehend their usefulness for us but to witness the miraculous *reality* and the unfolding *aliveness* of everything, a being connected together of all real entities. It is to encounter (rather than merely think about) the startling mutual dependence and interconnected significance of everything, the awesome awareness that all things live and have a purpose that is inextricably tied up with everything else.

Such an experience, then, is really to see life in a different way: from this point of view, life is not simply made up of "things" that just happen, as a matter of fact, to be, nor are those things put there just for our use and enjoyment. Rather, when you "see" through and beyond the individual things, you experience them against the larger backdrop of the miraculous and interconnected *life* of which they are a part! Life is a continuous flowing and unfolding, a startling upwelling and fantastic blooming, a sort of magical and ever erupting fountain from which life in its myriad forms flows. This awareness of the mysterious and interconnected ground of being that is manifested in the eruption of each remarkable thing is the experience of the sacred. And to encounter life in that way is what Native American life is all about. Once again, it is an experience, not a hypothesis or theory; and once you have it, you will never be the same again.

Soul Building

Why is that? Why does the experience of the interconnected sacredness of everything change you? And in what ways does it change you?

As I have tried to say, to experience the sacredness of everything is to see each thing as an amazing instance of ever flowing life. And to have that experience is to perceive oneself as likewise a miraculous instance of that mysterious life. To see the interconnection of all of life is to become connected oneself. Thus, you too must let go, must be open to grow, learn, and change. To see everything (and oneself as well) from this perspective, then, means that one will live differently. If life is an ever-flowing fountain, then I too—when I let myself be aware of it—am such a learning, growing, and changing entity, a process within the larger process called sacred life.

Of course, it is not just Native American cultures that are aware of this sacred interconnection. At one time, we were given the medicine wheel, to be guardians of it until the time that other people would be willing to do their sacred work. But of course, other cultures have similar teachings about perceiving the interconnection of all things and about becoming or living in connected ways. Thus, we can all learn from one another.

And what is it that we can all learn from our spiritual traditions? To become connected is to build your soul. It all begins here with the experience of the sacred. The heart teaches us that we are not just "we" but "WE" when we speak. Thus, our spiritual work means letting go of the separate and isolated ego in favor of a self that is connected to the whole. And such connecting is a whole way of life, a way of being, the good and real path in life—the good red path as we call it.

To become so connected, you must be open to the experience and not judge it. Let it be, and you will see the connection. That means you must accept it, but such acceptance can come about only by avoiding judgments and criticisms. Do not judge or criticize yourself or anyone or anything else for what one must do to grow into one's potential and purpose in life, for such judging and criticism separates you from what is judged and criticized; and the point, of course, is to get connected. So you must let go of that part of you that wants to hold back from the interconnection.

Getting connected, then, means to develop your soul. Soul building is what spiritual life is all about. In fact, for us, stones, plants, trees, and animals all have souls, but we don't! Human beings have to *develop* a soul because to have a soul means to know what your purpose in life is, how to fit into life. Trees and stones don't have that problem. They know their purpose, but we don't. We develop our souls by becoming aware of the interconnection of all life and then fitting ourselves into it.

Our traditional teachings are alive. If they were not able to adapt and change in order to help us move with the changes in life, they would become merely legends or myths, or "what was." They would become dead in the sense that they would not lead to real life. They would just fill the air. But that is not the case. Our spirituality has constant movement every day of our lives, as is appropriate in the realm of spirit. Because of what it connotes, we do not say that we have a "religion." Rather, we say that we have a philosophy, a way of life. It is a way of living in harmony with all things.

Getting connected, however, doesn't mean achieving perfection or having no problems. Certainly not. But it does mean living in a connected way, letting go of what separates you from the rest of life, accepting the creative surge of life within and all about you in a way that permits your true purpose or soul to emerge. That's why the drum is really our bible. Drumming reveals the interconnectedness of everything; it permits us to see ourselves, like the individual drumbeats, as part of an emerging whole. We encounter life in that drumming, and thus we can let our own souls emerge over against the interconnectedness of which it is a part.

Love Others

To perceive the sacred, then, is to see the interconnectedness of all of life. And to see that sacred ground of life is to grow and live differently from before; it is to develop one's soul. But what else does it mean? It means to be open to and love others—all the entities and people to which we are connected.

In this sense our teaching is not so different from the teachings of Gurdjieff, the great student of Sufi mysticism. The first thing any of us should do on the road to this unconditional love of others is, for example, to love plants. If we can love a plant without judgment, unconditionally, being open to it and permitting it just to be whatever it is, we'll have taken a real first step. Then, of course, we must move on to accept and love other entities, such as animals.

That is why the clan system stayed with many of us, because it is a rooted connection. Such systems are more than simply an animal symbol or representation of the clan. Rather, they manifest the rooted connections we have to the foundation of the universe. When we learn to love an animal, to accept it for what it is, then and only then can we even pretend to be on the road to spiritual understanding of ourselves. Only then can we begin to work together in the connected harmony that is reality.

I don't believe in the separation of things. In fact, that's one of the biggest problems we have. Think of all the separating and criticizing that goes on when we categorize people as "Native American" or "White" or "Black" or whatever. Instead of acceptance and becoming connected, we all too often separate ourselves and thus become disconnected.

Love all of creation, every worm and tree and drop of water. Love every entity just as it is. Love the sunrise, the eagle, each leaf of every tree. Accept and love everything for the wondrous miracle that it is. Take time to give thanks to the mountain before you climb it; open yourself to it and give thanks for its bounteous gifts.

Teach Your Children

To be so interconnected and to love others also entails that you teach your children this way of being, that you teach them that they too are interconnected to the wider community in which they live.

Some people have the attitude that if a child is not biologically related, they have no responsibility for it. One of the hardest things for me to learn has been that we are all part of one community and thus are responsible for everyone and everything else. That means that we all play a part in the development of children. In our culture, sensing the sacred connectedness of all of life leads to bringing the children into that community. So everybody has a responsibility to help raise children, particularly after they have been weaned. It's not just that they are told about this; they are actually brought into the community as a whole so that they come to acknowledge their responsibility to the whole. To be interconnected is to teach your children that they are interconnected too, to teach them that they must take their place within the larger cultural and natural communities of which they are a part.

Such community goes beyond the cultural teachings that have rooted us to this earth mother that sustains us and gives us a solid foundation for life. It goes beyond that. In my travels and conversations with the spiritual elders of many different cultures, I have found that they all, in fact, share this vision of life as a community that founds and supports a true life, what (as I have said) we call "the good red road."

Learning from Others

To realize this interconnectedness and community also means to be open to the particular ways other traditions express their spiritual vision. Native people, for example, have no problem with Christianity, meaning the teachings of Christ. Christ brought words of truth. But we do have a problem when Christianity becomes "churchianity," for then a kind of absolute and exclusive truth is claimed, which separates the members of that church from everyone for the purpose of subordination and control.

We have been sent out by our elders to learn from the teachings of others, not to become subservient but to hear and acknowledge the spiritual truth of connectedness being expressed by them. I have not found one Native person who refused to go and sit in the church of any other religion or faith. Here's the point. Our elders felt that we needed to be aware of how things were being expressed from other people's perspectives if we are ever to find common ground and become viable members of the larger community.

All too often, unfortunately, we have separated ourselves from one another because of resentment and fear and an inability to love and be open to others. We need to build bridges so that we can come together. We need to listen to other paths so that the interconnection and real community between us can occur. To do that, we must let go of the conceptions, theories, and limitations of our own path long enough to be open to other ways.

Our teachings declare that we have common ground and that the different religions are simply different paths to the center. In our words, the medicine wheel has four directions. We have different paths, but there is at the center of the wheel a fifth dimension, the spirit. It is there at the spiritual center that all the different paths come together and are interconnected.

The Drum

The drum is our bible, for it reveals the interconnective web or community that is life. It is through the drum that we are able to connect to the universal rhythm. In a real sense, then, the drum is a kind of summation of what I have been trying to get at here.

The drum was a gift from the women to the men of the tribe. In their ability to give birth and to nurture children, women embody the natural rhythm of life. Since men are not so fortunate, we are told, women gave men the drum in order that they too might connect to the rhythm of life.

Briefly, the drum is circular but never a perfect circle because only the source of the rhythm of life is perfect. At any rate, the circularity of the drum reminds us of the sacredness of life, that life is itself circular and cyclical, that no matter when I start in life, if I remember that life is a path, a circular movement, I can always come back to the starting place a better being.

The hide that covers the drum reminds us of our connection to the animal kingdom. It reminds us that animals provide shelter and food for us and that without them we could not be.

The rim of the drum is made of cedar wood. This wood is fundamental to the drum, and it reminds us that we and the rest of nature depend on the plant kingdom, that without plants neither animals nor humans could exist.

The playing of the drum, of course, likewise manifests the entire connectedness of everything, for the total rhythm could not be without each beat of the drum, nor could each drumbeat be without each and every other beat within the overall rhythm. So just as the drum reveals this whole flowing interconnection of dependent parts, we too come to see that we are part of a wider community in which we have a responsibility to the sacred whole and to all the parts of it to which we are connected.

Composting Our Connections
Toward a Spirituality of Relation

A moment of joy as unadulterated as I have known: sweating and panting, after gathering and layering dry leaves and green ones, earth and organic refuse, we began jumping up and down to mash the layers together. We took turns, singing and laughing, the Salvadoran children shrieking with hilarity as they bounced, finally grabbing hands of any age, sex, race, and nationality and spinning around until dizzy and done. This was the dance of the compost heap.

It seemed in retrospect that all my "issues" here joined sensuously in this local choreography of global healing: our guide Marta Benevides, the activist behind the project, herself a *feminist*, always working to empower other *women* in a *machista* society; linking our international, multiracial group of Drew University theological students with *indigenous* peasants with whom she is working to reclaim a bit of *land* and thereby to teach an *ecological* praxis for the sake of *social, economic, and medical* independence from the *neocolonial* superstructures of dependency. As a result of the war and the centuries of monoculture export agriculture, El Salvador suffers from the worst ecological degradation on the Latin American mainland. The people's ability to grow vegetables, fruit, and healing herbs in order to achieve an affordable healthy life is an act of economic and cultural resistance. It dissents from the dependency-breeding "free trade" future proposed by the U.S. neoliberal Initiative for the Americas. The United States–supported war against the FMLN, which involved routine massacres of peasants, was also a war against the earth: we saw the barren mountains, napalmed along with villages as part of the army's antiguerilla policy of "draining the ocean to kill the fishes." The apocalyptic aftereffect tells the story of the neocolonial heritage of depopulation and despoliation begun five hundred years ago. The life that rebounds there with such irrepressible counter-apocalypse witnesses to what has been lost—and what may yet be gained. As

Marta has risked her life for years to demonstrate, ecological, political, and cultural regeneration have become inextricable. Her vision pits "quality of life" against the obscene northern obsession with a quantifiable "standard of living." She defines quality of life as "human realization within a healthy environment."

I have found myself in the months since our trip returning often to this moment on the compost heap. I circle the scene, I replay it, I feel again the rhythm of the dance, the weight of the heat, the liberation of the laughter, the bounce of the earth. It has become a parable. It hints at how life on earth, like the life of the earth itself, is willing to come forth again if only we will work together, across our most difficult differences, to recycle our history and our common ground. In Cacaopera we collected and sifted through layers of garbage partly buried in the ground—broken bottles, rusted cans, toxic batteries, shredded underpants, and old shoes. So our society can live with quality only if it will sort through the garbage of U.S. consumerism, that garbage even now spewing itself with ever greater force through the planet along the routes of free trade. Sort through the toxins of the northern European–based civilization's assumptions of its own superiority and therefore its right to dominate and exploit. Sort through the levels of ecclesiastical and theological and perhaps even scriptural chauvinism that have inspired and sanctified our empires. As we sort, we may be repulsed, we sweat, we become grimy—but we also wake from our numbness. Then we can distinguish between that which can never be composted, which must be eliminated from our history, and that which can be recycled, like the leaves and the household waste.

My parable may seem too foreign, too political, or even too secular to address my assigned task: an ecofeminist perspective on the environmental crisis and "God." However, this parable has hardly yet exhausted its potentiality! For, of course, part of the point is precisely that ecofeminism in the United States cannot responsibly pursue its tasks without consciousness of these larger connections. Indeed, the ecological movement itself functions only as a wing of northern dominance inasmuch as it fails to integrate itself with the agendas of social justice nationally and transnationally. "Social ecology" is one operative label for the attempt to combine ecological with socioeconomic critique. "Social ecology envisions a world in which basic human needs are fulfilled through an economic restructuring that is environmentally sustainable" (Carolyn Merchant, *Radical Ecology* [New York: Routledge, 1992], 153. But most social ecologists still remain anthropocentric, indeed androcentric, in their focus, therefore failing to challenge the deep structures of cultural dominance. Therefore, the spiritual depth and gender of ecofeminist consciousness can collaborate productively with social ecology. But feminism itself is in a process of widening the focus of its critique and its proposals. Feminism, at least as I read it, no longer understands itself as a single-issue project, operating from a monocausal view of history (i.e., "patriarchy" is the sole cause of evil in the world). Of course this is why such awkward compounds as "ecofeminism" have come into being: to express the complexity of vision that seeks to embrace a multilevel and multicausal network of domination and resistance. Of course, accepting the label of ecofeminism can function as a further narrowing of focus—as in "I only deal with feminism in relation to ecology and with ecology from a feminist point of view." But I mean it not to subdivide but to multiply both the problems and the possibilities that come within our purview. For the "eco" of "ecofeminism" means nothing if not a disciplined attention to the interconnections of life itself and therefore of our study and practice of life.

Finally, of course, to remain a feminist may entail the claim that the subjugation of women by certain groups of men works in tandem with all oppressions—or rather, that the manipulation of the environment and of human beings for the benefit of an elite required and intensified the control of women's bodies. But at this point in history, White feminist work knows itself inextricably bound up in a tangle of ambiguities regarding our own complicities in race, in class, in cultural privileges, and of course at the same time in practices (such as the use of my word processor at this moment, the dishwasher running downstairs, the central heating that made this foggy October morning cozy, et al.) that, for all their apparent modesty within my society, make me part of the average "first world" population, which has thirty times the impact on the environment of the poor of the third world (in terms of technology, resources, and waste).

But I began with the image of a joyous dance, not of guilt and confusion and the paralysis they produce. The compost heap offers a metaphoric answer to our situation as persons enmeshed in the smooth destructive processes of late-twentieth-century U.S. existence. We cannot escape from our own world. But we need not. We need rather to take account of where we are, who we are, who we are with, and what we are for and thus proceed with our own recycling of ourselves and our cultures. One does not then hope for purification, for any unambiguous alternative space, either within our lives or within which we may live. For instance, in my own little amateur compost heap at home in New Jersey, I am aware that because I don't feel I can afford to buy organic vegetables for the most part, my supermarket produce will bring with it its worst chemicals—that of the peels and cores—to rot in my heap. But I know that this is still going to create a better fertilizer than chemicals I could buy and that eventually, as I learn to grow more vegetables (this is my first year), a real organic vegetable patch could emerge, with ever cleaner wastes to recycle. And perhaps just as important is the minor spiritual discipline that comes of beginning to take responsibility for your own garbage.

But let's cycle deeper into the metaphor. Attention to wastes is precisely attention to the dimension of life that rots, that ferments, that decomposes. It is attention to the finitude that is life. In other words, it is not death as a dramatic opposite and boundary of life—or even, in the language of the Apostle Paul, its enemy—but the death dimension that cannot be expelled from life, that cannot be separated from life, that life cannot be free of. Waste is the constant eviction from organisms of what has been used and yet not used up, of what makes itself available to pass into another life form, of what dies into the future as the long, slow nutritional base of what will be.

Is not this composting presence of death in life precisely, after all, what patriarchal civilization seeks to purge? Is not Western philosophy, along with the theologies it shaped, not based precisely on its earliest origins in the Parmenidian victory of immutable Being over Heraclitus' primacy of Becoming? In other words, has not Western thought remained riveted for nearly three thousand years to the vision of an essentially deathless and changeless presence, an eternal present, which the wise or the faithful are called to maximize within the framework of their own beings?

They experience their own finitude, their limitations and their deaths, their impotencies and their desires, as manifestations of a fundamental fall from the world of immortal spirit. Christianity, emerging from a Judaism that had not yet sought to transcend flesh and time—only to widen their possibilities in the people's *shalom*—struggled for an authentic synthesis with the Platonism that pervaded discourse at the time (like postmodernism today). But the celibate male elite, which dictated the course of Christian thought and its binding conciliar decisions, were themselves riveted to the timeless order of a changeless Being, which they transferred authoritatively onto the Being of God.

And so we cannot escape consideration of God, after all: meditation on waste, indeed on the waste of wastes, dumps us inevitably into a heap of old theology. When characterizing the flesh, the passing materiality of this world, and the distinguishing mark of women, the metaphors of corruption, deterioration, and the dung heap drop thick and fast. The assaultive force of the sexism inherent in much early Christian theology suggests the ancient background: the attraction to a pure and unadulterated eternity represented a liberation not just from the earth but from earth-identified female powers—powers that had been spiritually symbolized in the ancient goddess. We do not know whether in Stone Age or Neolithic cultures there was a more egalitarian arrangement of powers; but it would seem that hunter-gatherer cultures and early horticulture, and to some extent early Neolithic agriculture, did vest in women a power mirrored in the iconography of goddesses (An example of this would be the culture of the Iroquois Confederacy.)

Certainly, goddesses like Astarte/Ishtar, Isis, and Athena appeared in early form surrounded by natural creatures, often snakes and birds, to suggest this kinship of femaleness and the nonhuman realm. This imagery still lingered in the ancient world, subject to revivals such as the Isis cult, and therefore posed an ongoing insult to the ascendancy of masculine power—divine and human—over the creation.

The great monotheisms did not invent either the domination of women or the reckless mastery of the earth (as is made clear in the Gilgamesh epic, in which the patriarchal heroes insult the goddess Ishtar as they set forth on their journey to win glory by destroying the spirit of the cedar mountain, Humbaba).

The male-privileging spiritual denigration of earthly things as "lower" ranges from, at best, an appreciation of the beauty of the creation as a mere means to the end of worshipping its divine Creator to an ideal of indifference to the things of this world. At worst, this religious tradition secularizes itself in modern exercises of "dominion over the earth" as "man's prerogative." Modern science presupposed the power of dominion as the basis for that "progress" that has had such clear implications for the Western treatment of the environment. But this regime of deathless being did not emanate as a pristine epiphany from the minds of overly abstract men. Rather, it seems to accompany and to strengthen, however unconsciously, the unfolding of the politics in which it is nested. This entailed inevitably, the politics of the development of city-states and then of empires, based on a strict ordering of life into hierarchies of gender and labor.

As even Augustine explicitly argues in *The City of God*, the social order ordained by God depends on the strict order of obedience within the family. He refers of course to the order of the paterfamilias, who controls women, children, servants, and slaves and thus provides the microcosmic building blocks and requisite disciplines for the success of the state. Such patriarchal order readily extended to the control of "barbarians" and other colonized peoples and in the modern period enabled unprecedented imposition of itself on wide reaches of the planet's ecologies and peoples.

What feminists have repeatedly and in multiple styles, perspectives, and particulars disclosed is simply this: all that becomes subordinated to the timeless ordering mind of the patriarch, who images in himself God the Father, carries at once the taint more female, more natural, and therefore closer to the chaotic processes of finitude and death.

The ecofeminist flip of the operative order is to say, "Yes! Thank you!" If women's experience and history have kept us closer to the processes of nurturance and nature, if our lunar rhythms make it harder for us to deny our own flesh and finitude, so much the better! Let us claim our difference. Postmodern feminism, like the egalitarian liberal feminism of an earlier period, warns, however, against any identification of the feminine with the earth and the body, suggesting that we will pay too high a price for romanticizing physicality and generating for ourselves some essential and nature-bound identity. To all of which we might now respond: indeed, the recuperation of nature and of woman will and must proceed hand in hand simply because they have been devalued and exploited in tandem by the same sexist order. However, to identify any "feminine nature" as closer to nature will surely backfire—indeed, the quest for an essential identity already sells us out to the perennial patriarchal quest for a timeless essence of our being. Moreover, if certain men are "separate from nature," it must be said that this is by virtue of their cultural constructions of themselves as part of a transcendent elite; and it must also be said that certain women of that same class—praised as true ladies and, by the Victorians, as ethereal, the "household angels"—surely experienced far less proximity to the processes of nurture and nutrition, degeneration and waste, than did their own female and male servants. Besides, how can we imagine that someone is "closer to nature" than another, without admitting that human beings basically reside outside nature and therefore enter into relationships of it? If, on the other hand, nature is a name, and a precarious, culturally freighted one at that, for the matrix of relationships in which life unfolds, then what could lie outside it? Surely, no creatures. And what of "the Creator"?

For ecofeminist reflection, or at least for my own spiritual sensibility, the language of "Creator" has, like that of "God," alienating resonances. In the ambiance of Christian civilization, "creation" does not evoke some cosmic icon of woman giving birth, with the sharing of substance, the messy mutual implication, the high valorization of materiality that any picture of the Creator as birth mother must evoke. It smacks rather of the "over and above"—the very mode of transcendence that, when internalized, places humans over and above great female and nature-associated tracts of creation. "Creator" suggests some ex nihilo process of production whereby an artisan makes a pot and therefore does with that pot, itself relatively lifeless, whatever he wishes. ("Will what is molded say to the one who molds it, 'Why have you made me like this? Has the potter no right over the clay . . . ?'" [Rom. 9:20–21]). The pot is not of his substance. The Creator could live without it more or less as before. (Of course, there are less functionalist understandings of artistic creation available, but they belong outside the monotheist framework, with its strict critique of image making as idolatry.) And even more alienated is the most foundational of the biblical images of creation: the Word that summons the world up out of the chaos; in other words, a completely immaterial procedure, a purely verbal expression of an intellectual volition, an image appropriate for the peoples of the Book and therefore helpless in the face of the material devastation of the creation by human creatures. Such a Creator commands, blusters, summons us to obedience, even to good stewardship and participation in the new creation—"the earth groans in the birthpangs" (Rom. 8:22). These ancient images have their own tradition of a justice that heals human and earth imbalances. I respect their potentialities profoundly. But altogether, the tradition has had its chance too long and done too poorly to be trusted to rectify the situation from within, that is, from within a crisis for which that tradition is itself as much a problem as a corrective.

Yet this time more than ever calls for spiritual sensitivities to the processes of creation, of the unfolding of material life on the earth. I would not want to do without the Jewish and Christian metaphors of the divinity of the source of all life. And certainly I would not know what to do without the guidance of spirit images. Nor would I want to diminish the potentiality of revived images of deity as Goddess, as Mother of Life, as Cosmic Womb. Yet these anthropocentrisms also need to be occasionally relieved, suffering as they do from the strain of political struggles. Perhaps we might think of the divine as itself a great recycling process, as indeed the very heart of the metamorphosis by which death gives into life. We might imagine God as the great recycling center in the universe. Alfred North Whitehead, the source of process theology (perhaps the most ecologically attuned of all schools of Christian theology), pictured God's "saving" work as sifting through the junkyard of history. Or Levi-Strauss's trope of the *bricoleur*, the one who picks through the castoffs and oddities, who putters and creates out of garbage heaps. But if "God" is recycler, "God" is also recycled. For the sacred matrix of life is not something separate and outside life; the spirit immanent in life is therefore also altered in the process of transformation. That spirit is who it is only in relationship to what everything else is. That is also true of each of us: we are—and so express our holiness.

Such a divine process of recycling suggests at the very least that to waste, indeed to waste our wastes, to disparage material life and therefore to destroy it, is to go against the grain of the universe. It suggests perhaps even more: that to work to save life rather than to waste it, to recycle the castoffs, human or nonhuman, into renewed and valuable life-forms is to do holy work. If this work is holy, then those doing it *matter*, their work matters, and the material world matters. And perhaps the grace with which to work—the grace *of* works, not the old Protestant grace *versus* works—can be trusted. In Greek, "trust" is *pistis*, usually translated "faith." But faith here will lie not in a paternalistic power that will rescue us from the results of human waste and redeem us for a disembodied eternity. Trust rather creates and responds to the quality of relationships in which we "live and move and have our being." For to recycle is precisely to recycle the matter of our past, and the past is constituted of relations—broken ones, violent ones, sustaining ones, promising ones, unpredictable ones, human and nonhuman ones.

This complex weave of relations has a chance to be trustworthy only insofar as we continuously recycle its fiber through our very being— knowing that what I am at this moment is nothing but a creative composition, a kind of bricolage, a postmodern collage with its own ironic beauties, wrought of the past into a present that already shapes the future. Nothing is *not* an act of recycling in this sense: the past is always being made over into a present seeded with the future. But those who cooperate with this radical relatedness, who do not resist the ephemeral processes by which "I" continually appear and disappear, reappearing at a different instance of space/time and therefore as an "other"—those will manifest the marks of the holy, the rhythms of the changer and the changed.

Only after the dirty work of sifting and sorting, of isolating the toxins and piling the garbage, of gathering the bits of green and digging the pit, is it possible to begin to lay the compost into ground—and finally to dance. We did not stay long enough to see what grew in this enriched soil. Nor can anyone yet predict the outcome of the peace process in El Salvador, let alone of the global process in which politics, economics, and ecology hold the future under a question mark. But if the divine is the recycling center at the heart of our reality, the work will prove to be not necessarily successful but intrinsically worthwhile and therefore will continue to generate the energy and the community and the resources needed to carry it on.

I have just learned, having already written most of this chapter, that the village people did sustain the compost pile and indeed that the family whose house lay next to the land have contributed most of the house as a cultural center, a kind of museum of local culture, thereby providing access to the garden as an educational site. The compost apparently produced astounding cabbages, tomatoes, and a precious batch of medicinal herbs. The relationship of our Drew group to the Salvadoran community organizers who led the project has now recycled itself—we are talking seriously about incorporating this sort of experience into our curriculum.

But we are all here, not there. We might imagine "here" to be New England in autumn. Let me ask that we direct our attention to the recycling process that is the very nature of fall—all around you are the fallen leaves, some comrades still dancing in the wind, entering into the compost pile that carpets the earth. The relation of relations, the scintillation of all sentiences—this rots gorgeously, with no regrets, along with all that has fallen. So may it be with your body and work.

Recent Recommended Readings in Ecofeminist Spirituality/Theology

Adams, Carol, ed. *Ecofeminism and the Sacred*. Notre Dame, Ind.: Crossroad, 1993.
McFague, Sallie. *The Body of God*. Minneapolis: Fortress, 1993.
Merchant, Carolyn. *Radical Ecology*. New York: Routledge, 1992.
Ruether, Rosemary. *God and Gaia*. New York: Harpers, 1992.
Shiva, Vandana. *Staying Alive: Women, Ecology and Development*. London: Zed Books, 1988.

Part IV | BROADENING THE SCOPE

Apprehend God in all things,
for God is in all things.

Every single creature is full of God
and is a book about God.

Every creature is a word of God.

If I spent enough time with the tiniest creature—
even a caterpillar—
I would never have to prepare a sermon. So full of God
is every creature.

—Meister Eckhart

It is perhaps impossible to broaden the scope of anything beyond that of the cosmos. And this is the breadth represented by environmental conservationist John Carroll, ecopsychologist Albert LaChance, and geologian Thomas Berry in the three chapters in this section.

John E. Carroll is professor of environmental conservation at the University of New Hampshire and has directed university environmental studies programs for more than twenty years. He has published extensively in international environmental diplomacy and policy and teaches ecological ethics and values. He is co-editor (with Albert LaChance) of *Embracing Earth: Catholic Approaches to Ecology*.

A significant broadening or expansion is called for in John Carroll's treatment of a true ecological ethic. Such an ethic must be rooted in a broad and far-reaching humility before the magnitude and the magnificence of the Creation. Given the glory of Creation and given our apparent willingness to destroy it and our apparent capability to do so, suggest a need for us to reflect on balance and on what is needed to restore the imbalance we have so obviously created. What is called for is not only behaving ourselves ecologically in the future or even a willingness to restore what we've damaged but, much more broadly, a willingness to accept the necessity to replenish. In doing this we may follow the Christian tenet of repaying seventy times seven, or we may turn east to eastern philosophical concepts of *yagna* (i.e., giving back not simply as much as but more than we take), *dana* (i.e., giving something of ourselves to future generations), and *tapas* (i.e., recognizing the need to replenish ourselves, in body, in spirit, in soul.) Underlying any willingness to apply these profound concepts to our relationship with the planet must be an ability, first, to see the sacred in all and, second, to therefore come to revere that sacred. Such will inevitably lead to significant reduction in our ability to do harm. Likewise, getting a handle on the sin of pride, about which we are warned in all the world's religions, will significantly reduce our otherwise uncontrollable and seemingly unlimited ability to do harm.

Albert LaChance, who teaches at the University of New Hampshire, is a counselor, cultural therapist, and ecopsychologist known for his book *Greenspirit: Twelve Steps to Ecological Spirituality* and his founding of the Greenspirit Institute. He is co-editor (with John Carroll) of *Embracing Earth: Catholic Approaches to Ecology* and is currently writing on the architecture or framework of the soul and sacred process ecopsychology.

Professor LaChance likewise broadens our horizon with his approach to the human body, soul, and spirit. Just as Thomas Berry focuses on the damage to the outer, the planet, from what he calls the autism or deep cultural pathology of the individual, so Albert LaChance, a student and protégé of Berry, offers us his explanation of what is going on in the inner, in what he calls the architecture of the soul, and the sacred processing of that soul, that inner, through sacred process ecopsychology. This is expansion, or broadening the scope, in a very big way. While Berry broadens us in both space and time, LaChance broadens us in necessary response to Berry's conclusion: if we suffer from a cultural pathology or autism, how deep must we go within to find or to root out that pathology, that autism? And this is and must be carried out in context, that is, in the broad context of our evolutionary history and of our relationship to all other creatures in the Creation.

Thomas Berry is a Passionist priest, an octogenarian cultural historian, a self-styled "geologian," or theologian of the earth, and, through his books *Dream of the Earth* and (with Brian Swimme) *The Universe Story*, a leading spokesperson for the telling of the "new story" of who we are, where we've come from, and where we're going.

Thomas Berry is today widely known for the telling of a new story, the "universe story" as he calls it. His is a story that humbles the listener, the reader, for its breadth as well as its depth are almost beyond comprehension when measured against the "old story" with which we've been living for many centuries. It is the contention of these writers that the old story is now dysfunctional and no longer serves our purposes or our need for survival. Berry's is a story very much in line with ecological thought and with all that modern science has been conveying to us in these post-Einsteinian decades of quantum physics, quantum mechanics, and the development of chaos theory in mathematics. The old story that is dying is a product of Cartesian and Newtonian thinking and no longer has a place, given the new scientific reality and the vast adjustment in human perception it is causing. Berry contends that it is only by rejecting that old dysfunctional story of our separation from nature and our resulting hubris and arrogance and by accepting the new story, the ecological story, in humility and reverence that we can save ourselves. Such requires moving from what Berry calls the terminal Cenozoic to an emerging Ecozoic era, the latter defined as that period when humans would be present to the earth in a mutually enhancing manner.

All of these authors would argue that we must significantly broaden the scope of our consideration of the environmental problem, to begin to identify it as *the* human problem, if we are to get through it, to overcome it, to survive. The present narrower view will not suffice.

11 | JOHN E. CARROLL

On Balance, Replenishment, and an Ecological Ethic That Works

A colleague of mine often says that one of the things he feels is most missing from modern life is something called a "jaw-dropping experience," an experience so powerful as to cause one to react by dropping one's jaw and simultaneously making an exclamation, an exclamatory sound of real surprise. This is another way of saying that what is missing from contemporary life is a sense of awe and as well, the inspiration that comes with it. We really do feel we know it all or at least know all we need to know. Thus, there is little or nothing to be in awe about and little to inspire us.

Not being in awe of anything, it is difficult to feel a sense of the sacred, to feel reverence for anything, to treat anything as worthy of something as serious as reverence, to regard anything as sacred. I suggest the two go hand-in-hand, that lack of awe yields lack of inspiration and thereby lack of sacrality and the reverence that goes with it. Yet it is this missing element of reverence and its accompanying sense of the sacred that are missing pieces in the achievement of success in overcoming environmental problems.

So how do we achieve or inspire them? A big start, I suggest, is in understanding them and focusing on them.

A first step concerns pride, or hubris, and recognition of its presence. A basic reason that the Amish communities of the United States send out multiple families (often seven large families) when they wish to colonize new land or start a new settlement relates to their strongly held belief, as a Christian people, that pride is the greatest of all sins, the true root of all evil, perhaps even humanity's original sin, and that human beings acting alone do not have the means to overcome it. Indeed, in their view, the belief that individual humans can alone overcome the force of pride is held not only to be arrogant but to be itself an example of the sin of pride. The

Amish believe that humanity obtains the power needed to overcome pride or hubris only *in community* and with the peer pressure that community brings about.

As a non-Amish and secular people, we not only see little or no need for community (beyond paying lip service to the word) but in fact champion an individualist notion that is quite the opposite. And when we speak of pride as a problem, it is something we call "false pride" or "overweening pride" that we identify as the problem. This enables us arbitrarily to assign the problem to a question of degree, an arbitrariness that further enables us to avoid the problem and to justify a level of arrogance toward nature that Christian (and other) religious tenets really do not permit.

Pride gets in the way, as the Amish people know so well. Pride (hubris) inhibits a sense of awe and prevents the jaw-dropping experience from occurring, since both awe and that jaw-dropping experience are based on pride's opposite, humility—indeed, on a high degree of humility and a sort of informed naïveté. A sense of awe and the jaw-dropping experience, the exclamatory experience of recognizing nature and the system of which we are a part, should be natural. They should come naturally, as they do in young children. One does not (indeed, cannot) create them, although one can create the circumstances necessary for achieving them.

E. O. Wilson, the Harvard biologist, in his exhaustive work on life's great diversity, gives us a realization of how much remains unknown, of how little is really known, about life, about the world. We are as sixteenth-century explorers, he tells us, and we don't know it. (In all probability we don't know it because we don't want to know it.) We know, Wilson tells us, less than 10 percent of the species to be known, and it is quite possible that that percentage will even drop, not, obviously, because we'll become aware of more species but because, in the search for more species, we'll become aware that we are unaware of many more. We are on the verge of discovery, Wilson tells us. Indeed, we always are, for the goal becomes more and more elusive.

And what does this realization lead to? It leads to a sense of humility, of respect, of awe, of reverence, of the recognition of sacrality when we see it—a reverence for the sacred. Therefore, it leads to an environmental ethic that really works.

We think of ourselves today as a people who know virtually all of the planet's rivers and mountain ranges and most of its other physical features. We assume there is nothing left to be discovered. And yet it can be said that we don't know very much of the physical world either, for we have seen it only through the lens of Newtonian physics and Cartesian–Baconian science, not at all through the lens of twentieth-century findings in physical and ecological science, in mathematics or biology. We appear to have only one construct, one referent or reference point and seem to know almost nothing about any others.

A whole new world waits the removal of our blinders. Recognition of this would send us a long way down the path to a proper environmental ethic, one in keeping with ecological reality, one sure to give us the ecological sustainability we claim to so dearly desire. To maintain and protect that world, we need to consider balance and replenishment.

On Balance and Replenishment

Critics of environmental protection and environmental conservation measures, especially corporations and business, often argue for something they call "balance" or a "balanced approach" in land conservation and environmental protection. This implies a balanced relationship between, on the one hand, use, alteration, and destruction and, on the other, nonuse, protection as is, and preservation. It starts from the assumption that no use, alteration, or destruction has yet taken place, an obviously erroneous assumption and therefore the false assumption that there is an even playing field from which to start anew. This approach, of course, belies history and misleads to the extreme. An obvious case in point is the old-growth forests issue, wherein those advocating cutting and suggesting need for a balance between cutting and preservation choose to ignore the fact that a full 90 percent of all old-growth forests in the United States have already been cut, have already been destroyed. Ironically, the call for balance coming from cutting advocates should indicate no further cutting at all, together with very ambitious attempts at replenishment and regeneration of what has already been damaged. The arithmetic and the philosophy of true balance are fully on the side of the "no further cutting" advocates. This is almost always the case, regardless of the issue. Present calls for balance, given whom they are coming from and the philosophy they represent, are nonsensical.

What is truly needed, therefore, to move in the direction of balance is not more utilization and exploitation but restoration—restoration of destroyed, altered, or ecologically contaminated habitat—to use an even stronger word, which this writer regards as preferable, replenishment. The basic laws of ecology (so eloquently expressed by Barry Commoner some years ago as everything is connected to everything else; everything must go somewhere; nature knows best; and there's no such thing as a free lunch) give general guidance on how to behave ourselves ecologically, and they set the stage for restoration. But we must turn to a deeper way of thinking and knowing to achieve replenishment. We are given this guidance in three concepts from the East, each of which has numerous corollaries in our own Western thinking and philosophy. These concepts are *yagna, dana*, and *tapas. Yagna* refers to giving back more than we take, from the planet or from one another, in other words, replenishment manyfold. *Dana* refers to giving something of ourselves to future generations. And *tapas* refers to the need to replenish and renew ourselves, our spirit, our souls.

It would be a mistake to interpret *yagna* as "sacrifice" in the Christian or Western sense. The Christian use of the word *sacrifice* suggests payment, appeasement, or redemption. The Hindu and Jain term *yagna* starts and ends at different points. It starts at the ecological fact that, at the heart of all life, to live is to devour life. And it may be said that what is permanent is not the devoured or the devourer but rather the process of devouring. Likewise, the sun gives light by devouring itself. Fire burns and devours—to be kept alive it has to be fed fuel. Since devouring is the only permanent entity and thus more important than the devourer or the devoured, sacrifice becomes humanity's entrée to the cosmic process. It is our essential contribution to our own continual existence and to that of the cosmos. The law of karma says that every self-interested (i.e., nonsacrificial) action or work has consequences (whether good or bad) for our future life or lives. So the only way to escape the eternal cycle of rebirth and contribute to the continuity of the cosmos (and the narrower ecosystem) is to engage in a life of non-self-interested action or work, that is, a life of sacrifice. Sacrifice, in this broader sense, becomes necessary for survival, ecosystemic and otherwise. It is not an option.

Dana is self-control. It involves the reduction of wants but need not necessarily be viewed as sacrificial, and *tapas* involves spiritual fervor or ardor, including austerity, asceticism, and penance. Both involve concentration, focus, silence, and a degree of necessary solitude. *Dana* is at the center of all self-worth. We can find our home in this world when we find our purpose. And if our purpose is to nurture all life, we begin to fulfill our lives.[1]

What emerges from this is the totally ecological dictate, a simple moral of life, to consistently strive to live with less. We may wonder why we spend hours rushing around if the moral of life is that simple! Part of the answer may be that Western culture tries to make the moral so very difficult to comprehend that we spend our lives trying to comprehend it instead of living it. *Yagna, dana,* and *tapas* are all ways of living life. They are not morals to ponder but morals to live, and they lead to replenishment, a very different action and a very different result from that of replacement. Any Christian should readily realize that these are also central teachings of Jesus Christ, Christ the ecologist.

Rachel Carson, scientist and writer, when writing of the sea in her elegant book *The Sea Around Us,* gave us some perspective, some context in which to view this necessary replenishment, and a warning of what would happen if we stray from this task: "It is a curious situation that the sea, from which life first arose, should now be threatened by the activities of one form of that life. But the sea, though changed in a sinister way, will continue to exist; the threat is rather to life itself."[2]

Or as scientist James Lovelock has told us, we need not worry about saving the planet. The planet can take care of itself, with or without us. It is we whom we need to save, for the planet, perhaps even in its own self-defense, may perchance need to eliminate us or at least to remove the conditions necessary for our continued existence. If we don't, therefore, adopt the replenishment mode and at least variants on *yagna, dana,* and *tapas,* then we will surely have ourselves and our very survival as a species to worry about. Imbalance will not be tolerated, either in the ecosystem or in the greater cosmological system. Our job, therefore, is to restore the ecological balance through the work of replenishment. To answer the ecological challenge before us, we need to cut a path that we can follow.

Answering the Ecological Question: A Threefold Path

If one accepts the growing notion that the ecological question, as a challenge facing humankind, is fundamentally a philosophical question, an ethics and values question rather than a scientific, a technical, an economic, or a political question, then one might well posit an answer to the ecological question as a threefold path.

Buddhist philosophy and spiritual practice teaches that there is an eightfold path to right livelihood, to solving the dilemmas and answering the challenges that life presents to us. This is somewhat akin to Judeo-Christian instruction and guidance presented by the Ten Commandments and, in a more positive vein to Christians, the Sermon on the Mount.

The modern science of ecology instructs us in certain realities pertaining to the interrelationship and interdependence of all life (and life is broadly interpreted to include the inorganic as well as the organic). These realities

posit a specific life-style and philosophical attitude governing daily existence, governing how we are to be in the world.

It is reasonable to conclude from the common core of philosophical,
religious, and spiritual interpretation and from the core of the findings of
ecological and physical science at this time near the end of the twentieth
century that the answer to the ecological question could well be posited as
a threefold path. This path would take the form of reverence of (and not
simply respect for, which is a lesser charge) the world around us, with
which we live in relationship and have dependence on; of an approach to
that world that recognizes the dependency relationship, suggesting a technology and a practical method of living that is appropriate; and a studied or
reflective consciousness of the universal human weakness of overweening
pride or hubris toward that world and our relationship with it.

Reverence

We often encounter the notion that we are to have respect for life, for
nature, for the natural system. This is not something we have generally
carried out in practice, but it is indeed put forth as an ideal. And yet there
is something critical lacking in this notion, as it is too readily trivialized in
our relationship to the cosmos. The idea lacks the deeper and much more
important connotation of reverence. And why is this so? It is because that
for which we claim it important to have respect we do not consider (or no
longer consider, as we once did) to be sacred. We have no notion of the
sacred, of that which is sacred, and thus we find it most difficult, nay
meaningless, to engage in reverence. We can respect many things, but, I
submit, we cannot have reverence for, we cannot revere, what we do not
consider to be sacred. And what is "sacred"? What does it mean to be
"sacred"? It has been my experience, with the limited knowledge I have of
non-Western philosophy and religion (both Eastern and indigenous
peoples' thinking) that the word *sacred* is applied in many, many circumstances within those philosophies and cultures, much more than in our
own lives, and yet the word continues to be meaningful (for practitioners
of that thinking) — it does not become trivial. For example, Hindu, Jain,
and other Eastern cultures in India tell us that the Ganges is a sacred river.
To me, this means it is considered both special and unique relative to other
rivers. Perhaps it is special, but unique, as we westerners would use the
term, no, for one finds that many other rivers in India are also considered
sacred, so much so, in fact, that the Western mind readily forms the idea
that all rivers, at least all rivers known to and used by humans (and perhaps

those not so known or used) are sacred, that there is no limit to sacrality here. One finds much the same situation in Native American thinking. And one can find the same thinking with respect to native old-growth forests, to animals, to rocks, and more.

My Western mind suggests that this is trivialization and is ultimately meaningless. But is it? Are not all rivers, is not all water, necessary to human life, necessary to all life, necessary to the ecological pattern and indeed to physical "laws"? Sufficient thought, sufficient contemplation, can lead even a Western scientific rationalist, a Cartesian–Newtonian thinker of the highest order, to think in such a manner and to see behind triviality and meaninglessness to a deeper meaning. Can each river be sacred? Yes. Can one river be more sacred than another? Perhaps not ultimately. Can a blade of grass be sacred? Yes. It has taken this scientifically trained Westerner many years to realize that no one river is more sacred than another but that they are all sacred, not simply collectively but each in its own individual way. They are sacred to ecological pattern and sacred to humanity's life on earth, sacred to the continuing functioning and development and evolution of the cosmos on its evolutionary journey.

And where does "reverencing" (if I might use that word) enter the picture? It is obvious to all that human beings, through their great brain development and mental/intellectual ability, have the capacity, the means, to destroy themselves, to destroy much of life on earth, to alter much of the established ecological pattern, and likely even to make the planet uninhabitable for forms of life akin to their own. (There is doubt as to whether we have the ability to make the planet unsuitable for all life, including bacterial and other microscopic forms; it is perhaps the ultimate in arrogance to believe we could do so.)

Then there is the matter of the effect of reverencing on us, on the human psyche. It was high in the Colorado Rockies in the company of a Native American spiritual elder that I first gained appreciation of the notion that our ability to alter, change, destroy, or manipulate the natural system for questionable ends was weakened in some proportion to our identification with and reverence (again, not respect but reverence) for that natural system and its various component parts, organic and inorganic. Indigenous American notions of the "stone people" (i.e., of imaging stones and rocks in this way), as well as other well-known indigenous peoples' tales of relationship and ways of being with and, more important, in (i.e., as a part of a larger cosmic whole), are further support for an ability to feel a part of that larger whole. In consequence, knowing deliberate damage to any part of that whole becomes knowing deliberate damage to all other parts. As we come to know that self-destructive behavior toward any one part of our own physical being can hurt all other parts, we learn that physical damage to our broader being—the ecosystem, the component parts of the ecosystem, and, most important, the web, the pattern of relationship that binds us all together—hurts us in whole or in part. Such experienced knowledge can, I suggest, significantly reduce our ability to destroy, to damage the pattern, providing an important restraint in our own species' long-range best interests.

Ecologist and ecological philosopher Edward Goldsmith, editor of *The Ecologist*, has said that the day humanity stops destroying the environment is the day humanity loses its ability to destroy, its means to destroy, and not before. In my narrow way, I once took this to mean the loss of economic ability through collapse of the economic system or perhaps general war, anarchy, or ecological catastrophe. But perhaps gaining a sense of the sacred and refining an ability to "reverence" the sacred, to revere the pattern, is the way that limitation, that apparently necessary restraint, can be achieved.

Technology That Is Appropriate

We owe to British economist and philosopher E. F. Schumacher much of our understanding of appropriate-scale or intermediate-scale technology, a technology that Schumacher often called "humanscale." Contrary to the thinking of some, Schumacher was not so much a critic of technology as he was a celebrant of technology. But in his celebration of technology he helped us see the value and beauty of efficiency, not the sham form of efficiency we often encounter but true efficiency in the best rationalistic approach to the concept, namely, the best balance of inputs and outputs answering the question "To what end?" and thereby ensuring appropriateness. To the person who would ask, "Appropriate to whom or what?" Schumacher would answer, "Appropriate to the task at hand." Thus, Schumacher's idea of appropriate technology is not subjective at all; it is objective in the strictest sense and rational in the highest sense of Western scientific rationality. It is always measurable to the task at hand and internalizes all externalities. It is both rational and efficient and by its very existence argues that the prevailing technologies of our society are neither rational nor efficient. And it is intermediate in scale in that it lies between the truly primitive technology of ignorance and the so-called high technology and fossil fuel–based and high energy–driven technologies, with their many costly externalities and environmental and social consequences—the

technologies that Schumacher viewed as overkill and that Dennis Meadows and others involved in the "limits to growth" argument describe as
"overshoot," a shooting so far past the mark that we now find ourselves in
deep trouble. And Schumacher, who is closely associated with the concept
of "small is beautiful," finds that the world in which he (and we) reside is a
world of too large scale, of too much centralization, an inhuman-scale
world rather than a human-scale one, which has gone beyond humanity's
ability even to conceptualize, never mind control or make peace with. So,
much of intermediate technology focuses on relatively smaller, decentral-
ized systems controllable under the hand of individuals or small groups
of humans, placing humanity as master of the technology rather than slave
to it.

The "reverencing" described earlier naturally leads to a reverence for
efficiency, not as the word is all too commonly used, an incomplete and
incorrect usage, but in its true meaning and correct usage, considering both
the input/output energy ratio and suitability to the task at hand. It leads to
a respect and an admiration for technology, for technological systems as
efficiently constituted, and for materialism in the best sense: a high regard
for good materials, for a high level of design and craftsmanship, for long
lastingness, for repairability, and for such materials' ability to perform a job
simply and well with a minimum of ecological and social cost. And of
course, such materials and system must yield confidence and self-esteem to
the user as human being, rather than reducing and belittling the user as is
characteristic of so much technology today. Appropriate technology is thus
a critical and important tool, a necessary vehicle, on the threefold path.

Overweening Pride or Hubris

The third part of the threefold path does not involve eliminating false pride
or hubris, for that is not humanly possible. The Christian Bible tells us this.
The life and work of the Amish people of the United States is a living
expression of this reality. Indigenous people tell us this. Eastern philosophy
in various and sundry ways tells us this. What is involved here, however, is
awareness, as extensive and as deep as possible, and an accompanying
desire to do something about reducing it. The latter act goes far to address
the question, which is likely all we can hope for or aspire to.

Christian religion has long held the sin of pride as a critical and central
tenet (although recognition of this is probably less in Christianity today
than in the past). It is a sin, a violation, that runs through all the central
dicta within the Ten Commandments and, as well, the Sermon on the

Mount, but it is more subtle than any of the other elements of those dicta. By referring to it as "false pride" or "overweening pride" or "excess pride," it has been possible for humanity to put it aside relatively more easily than might otherwise have been the case, especially collectively during the past three centuries or more of Cartesian- and Baconian-derived superiority over nature held by at first Western and ultimately almost all peoples of the industrially "developed" world. Perhaps because of the degree of comfort taken by the use of qualifying adjectives like *excess, false,* or the less common *overweening,* the word *hubris,* which doesn't allow for such comforting qualifiers, has gone out of use. It is, however, returning now to more common usage and is much promoted by American ecological/agricultural scientist and philosopher Wes Jackson in his many writings and presentations and as well by philosopher/poet Wendell Berry, among others. Biologist David Ehrenfeld has referred to it only slightly differently in his landmark book *The Arrogance of Humanism.* It is arrogance, indeed, and the kind that can get us into no less serious trouble than that encountered by the characters in the biblical stories as told over millennia. Indigenous American thinking, from the famous letter of Chief Seattle on through countless cautions and warnings we have received from such people so close to the earth and its rhythms ever since our contact with them, consistently warns us, too, of the latent dangers of false pride.

The Amish of America, in their intense conviction and behavior concerning community, are living examples of the problem. It is instructive that the Amish, such strong believers in and practitioners of community, perhaps the strongest in the Americas, see the vital nature of recognizing this problem and the inherent challenge it presents. When the Amish set out to colonize a new area, they generally send out seven families so that community exists from the start. Since Amish families are characteristically large, this can involve thirty-five to as many as fifty-six individuals. The Amish, as Christians, believe that life in community is absolutely necessary to conquer the temptation to false pride that lies within every individual human and that no individual human, they believe, is capable of overcoming alone. Indeed, they believe it is a sin of false pride itself to have the arrogance to believe that the sin, the weakness of pride, can be overcome by any individual human. It is the task of individuals to try, but only in community can the task be achieved. Hence, the Amish not only live, and live intensively, in community, but they only colonize new territory and establish new farms in community as well, preparing the path for the large community that is to follow. False pride, the arrogance to think of oneself, the individual, as greater than the whole, is the original sin of Adam and

Eve, the original sin, if you will, with which we were all branded and not only in the Christian viewpoint. It is the sin from which we as humans cannot escape singly. It is the sin that has gone far to help us damage that ecological pattern described earlier. Consideration of such is not only appropriate for treatises on religion but also for treatises on ecology and on how we are to be with nature, with the pattern of which we are a part, as well as with one another.

Perhaps the notion of a threefold path is too simplistic. Even the Buddha gave us eight. But maybe it's a start in the right direction or, in any event, an important direction. And we do need to start that process today. In this way we can secure for ourselves an ecological ethic that really works.

Notes

1. Gratitude is expressed to Richard Forsythe of Brighton, England, for sharing his insight on *yagna, dana,* and *tapas.*

2. Rachel L. Carson, *The Sea Around Us* (New York: Oxford University Press, 1951).

The Architecture of the Soul
Sacred Process Ecopsychology

We just cannot go on this way. The reality we've created is more than we can bear as individuals, as cultures, as a species. Much of the time we are forced to block it out, choosing some form of unreality. The energy needed to face the monstrosity of cultural disintegration robs us of our vital energies, and we often tumble into depression, despair, and indifference. To live in unreality, in denial, is to avoid the pain temporarily, yes, but we cannot solve the problems of reality from that vantage point. There is no creative energy in escape, and this loss of creative energy is the greatest energy crisis we face. To live for long in unreality is to succumb to certain emotional and mental illness.

Horrendous levels of violence and distortion assault our emotions daily. Another woman has been butchered along with her children. Another president, senator, or Supreme Court justice has been caught peddling the big lie. Another liar has been dragged, kicking, screaming, and denying, from office. Another tanker has split open. They're going to build a copper mine in the Tatchenchini River Valley in Alaska. The frogs are dying globally. Families are in tatters. Genocides, bombings, rapes, starvation, molestations, multiple millions of abortions, ghoulish experiments legalized on the victims of abortion—the list goes on and on. The frayed legal establishment cannot for long contain the explosion. Something is very wrong. We just cannot go on this way.

In the past our religious establishment would have offered guidance at such a time of crisis. Too often staffed by the least creative among us, lacking in true spiritual authority, scrambling for language and techniques that appear to have relevance, the religious establishments themselves suffer from a collective bipolar disorder. On the right we have the bake-sale goody-goodiness and those who decide who is saved and unsaved as though the Most High God were a sort of cosmic Joseph Mengele. Quasi-facist undertones pervade the religious right. Good evangelists like Billy Graham are too often drowned out by the money-grubbing phonies of television fame. On the left we find a sneering one-upmanship, self-righteous and convinced of its superiority over the traditionalists. Here are the Christian New Age types, "keeping it positive," so sensitive to everything feminist that they support in utero genocide while labeling the rightists "christo-facists"! If the above seems simplistic, it is because in the extremes things are simple, black and white.

Many of us are uncomfortable with either of these poles. Many of us have opted for religious disfranchisement rather than identifying with either group. But when the centrists are gone, the polarization only gets worse. The arthritis of the right becomes even more rigid, frigid, shallow, and restrictive. The unboundaried cancer of the left grows still more amoral. Many of us love our Jewish, Catholic, Orthodox, and Protestant traditions—and we are committed ecological activists. We are alienated from both poles of the pathology. We want faith, morality, tradition, and a vital planet. We want life, and we want to praise the Author of life. We need a third option.

Not only are human cultural systems collapsing, but the planetary life systems in which they are nested disintegrate all around us. We must deny this knowledge in order to survive and function day by day. Still, we know it. There is a fundamental linkage between violence, religious and moral bankruptcy, epidemic insanity, in utero and ex utero genocides, species extinction, and the disintegration of the natural world. The inability to form sustainable families is related to the extinction of the primates and mammals. The violence that men do to women and children and to each other is linked to a violent relationship with life itself. All of our seemingly insurmountable problems are in fact many tentacles of a single cancer. The deathliness coming at us from all sides is just a boomerang of the deathliness we deal out to all sides. The horrors we face are mirrors of the horrors we dispense. For every action there is an equal and opposite reaction— Karma. As we sow, we reap. The measure we measure is measured back to us. If we stop running from responsibility, we can name and deal with our problems.

What then do we call this disease? How can we diagnose it in order to deal with it at its root? I'll answer this question later in the essay. But to begin with, we need a new way in which to see ourselves. We need a comprehensive model of what it means to be a human person, one that includes that which has proved useful from the past and that which we have lately come to know. This new model must be, at one and the same time, psychological, spiritual, cultural, biological, ecological, and cosmic. In my book *Greenspirit*, I outlined some of the ways in which we could begin to do this work. I placed them in a twelve-step format in order to make them available for everyone. In my second book, with John Carroll, *Embracing Earth*, we gathered many insights from Catholic writers that might help us to understand our dilemma within the context of one Western tradition. In this new project, "The Architecture of the Soul," I shall attempt a personality model that will help us to see ourselves in this new way.

Sacred Process Ecopsychology

Following C. G. Jung, I am holding that the psyche has three basic zones. I have named these zones somewhat differently and have nuanced them into more basic and specific subzones or layers and phases. I have listed them below.
Zone One: Consciousness
    Subzone 1: Cognition
    Subzone 2: Affectivity
    Subzone 3: Instinct

Zone Two: Subconsciousness
    Subzone 1: Personal
    Subzone 2: Familial
    Subzone 3: Cultural

Zone Three: Preconsciousness
    Subzone 1: The Primal
    Subzone 2: The Mammalian
    Subzone 3: The Biogenetic
    Subzone 4: The Geogenetic
    Subzone 5: The Cosmogenetic
        Phase one: The Explicate (Immanence)
        Phase two: The Implicate (transcendence)

In this brief essay I will give a simple description of each of the zones, subzones, and phases I have articulated above. In each case I shall include a "physical corollary." My reason for doing so is to ground this model in material reality. At all levels the universe is both mind and matter.

Consciousness

Subzone 1: Cognition (Physical Corollary:
The Neomammalian Cortex)

This first subzone of human consciousness, cognition, is the seat of thought. Among the many personality theorists, I find the work of George Kelly of particular interest. Kelly holds that we come to know by creating larger and larger systems of thought. The following example might help. At whatever level we are aware in utero, that awareness is circumscribed by the watery habitat within which we are contained. The world beyond our mother's womb enters our awareness only through our muffled experience of sounds that reach us there. After the trauma of birth, our ex utero awareness would be the container of our mother's arms. From there, we might become aware that we are contained by our crib. In time we might become aware that our crib is only one piece of furniture contained by our bedroom. Later we become aware that our room is only one contained by the house. Next we realize that our house is only one in the neighborhood, the neighborhood only one in the town, the town only one in the state, the state only one in the nation. In each instance our knowledge is superseded by and contained within the nest of a larger knowledge. At each level the object known preceded our awareness of it and provides the architecture of our knowing.

There are larger superordinate constructs than the nation, though few people identify with them. Our nation is only one contained by the continent in which it is nested. Nations are only as healthy or as rich as the health and richness of the continent in which they are nested. The continent is only one nested in Earth. Earth is only one of the planets nested in the solar system, and the solar system is only one among many nested in the Milky Way. The Milky Way is only one of billions of galaxies nested in Cosmos. In this way we come to know more and more and contain our knowledge within larger and larger ordering thought systems or, as Kelly calls them, cognitive constructs.

Before passing from this sketch of cognition, it might be important to point out that if we are to survive as a life community, we must all move beyond the cognitive construct of the nation-state. At present we must all become species-conscious, continent-conscious, and planet-conscious. If we are to become fully human, we must also become Cosmos-conscious. In this way, both subconscious and preconscious zones of the psyche will ascend into our awareness. Humans being a functional phase of the cosmic process, will provide a space through which plant and animal life, the continents, and earth herself will rise up into thought and into human sensitivity. Correspondingly, it will become increasingly obvious to us that cosmos, earth, and all species provide the architectural supports for the human soul.

Subzone 2: Affectivity (Physical Corollary:
The Paleomammalian Cortex)

Feelings are the sensitivity of an organism. When the life process ends, sensitivity at the organism level evaporates. What remains is sensitivity at the molecular and atomic levels, as well as at the subatomic level, as the organism dissolves into its material components. We experience many levels of feeling, from the cellular to the tissue and organ level, to feelings that subsume our entire selves. There are feelings associated with the cognitive process noted above. When our understanding wraps around and penetrates a previously incomprehensible idea or problem, we feel elated. To lose a loved one is to experience a feeling response that involves our whole being, bones to soul. When we experience a revelation in spiritual knowing, we feel larger, deeper, more human, more divine. When we resolve a personal conflict from the past, we feel release. When we form a family, we feel more secure. To watch our family dissolve is to feel depressed and bitter. We feel pride in the achievements of our culture; we weep at the singing of a national anthem. We wonder at animals, at plants, at mountains, at rivers. We thrill to watch whales and other great mammals. We laugh at monkeys. Receiving flowers can spark feelings of love, forgiveness, and comfort. We are stunned to contemplate our origins when we gaze at the night sky or the ocean.

At present one of our most racking dilemmas is that we have tried to cut ourselves off from our feeling responses toward each other; toward each other's cultures and holy books; toward the animals, the plants, and insects; toward the mountains and rivers; toward the whole created order. A hardened and violent shallowness has been the result of this shutting down. When we cannot feel the pain of life, we go on inflicting it unawares. All around us the architectural supports of the human soul collapse, and we can't feel it. Religious fundamentalisms actually support us in this shutting down. Disassociating themselves from the larger family of life, they deny the great handiwork of the sensitive God who found it good. Often calling themselves "prolife," they claim the right to ruin the context for life if that destruction serves immediate human ends. On the left, claims are made to support the sanctity of nonhuman life and for the context of life, the habitat of earth. The one exception to this sanctity is human life and the habitat of the human womb. We need a third option. That option is sensitivity to all life.

Subzone 3: Instinct (Physical Corollary: The Brain Stem)

Intuitions are spontaneous experiences of reality. They derive from our own felt responses to what we experience within and outside us. To lack these feeling responses is to lack beliefs. Martin Luther King Jr held powerful beliefs because he felt powerful feelings, both about his own suffering and the sufferings of others. When we feel the agony of the nonhuman world and respond to those feelings, we become ecologically active. When we are no longer able to prop up the legalistic defense mechanisms of "prochoice" rhetoric and begin to feel the twisting agony of preborn children, dismembered alive at our whim, then we begin to have beliefs about the sanctity of human life. Feelings arise from the entire life community into human awareness if we allow them to. We begin to possess beliefs because we feel the truth of our experience.

We hear so much about boundaries these days, especially in counseling circles. Boundaries are the limitations we place around ourselves and our behaviors. These self-limitations are what many call morals. Morals are to humans what niche and instinct are to prehuman life forms. Niche is the way in which an organism fits into its physical locale. Niche is limited by the physical properties of habitat or locale. For instance, the seashore is the boundary that limits land animals to land and marine animals to water. These boundaries are deep physical and psychic structures associated with the nature of life itself. In human consciousness they are caught up into self-reflection. Humans self-create niches and then self-impose these limitations or boundaries on their own behavior. Our loss of boundaries around violence could very well be associated with the loss of boundary wisdom in nature resulting from the ruination of habitat and the niche of animals, birds, insects, and plants within that habitat. If we receive the news of evolution through the brain stem and that news arises from the prehuman zones of life, and if we are destroying those prehuman zones with their natural boundaries, then we are condemned to lose the ability to have boundaries ourselves. Neither therapy nor mood-elevating medications can create those boundaries. Only the combined effort and wisdom of the life community can do that job.

Subconsciousness

Subzone 1: Personal Subconscious (Physical Corollary:
Oneself/One's History)

Between those psychic contents to which we have immediate access in consciousness and those that are barely accessible at all in our preconscious minds, there lies this Hades of the psyche, the subconscious. We all have memories of things we've done or of things done to us that, for a variety of reasons, we'd rather not be conscious of ourselves and that we certainly hope to keep from public view. Bodily functions, secret pleasures and pains, and things that feel too big to deal with daily are stored here. The denied terrors associated with life in this century are stored here as well, living out their lives in migraines, addiction, and any number of neurotic symptoms, including lives lived in escape from reality. Abuse victims often bury their memories here. Lies, thefts, personal failings, and shames are stored and half-remembered here. They live and lurk just below consciousness like the shades in Hades.

Phobias: low self-esteem, chronic depression, and rage are some of the symptoms of their presence. When they become painful enough, we can begin to speak not of Hades but of hell. Leaving this zone unexamined, which most people do, leads to three very negative results. First, queasy feelings tend to leak "up" into consciousness like the breath from a sewer. These feelings tend to make us suspect, dislike, and even hate ourselves. Second, these same feelings tend to leak out beyond the boundary of the personality like a psychic leachate into our family systems, our work environments, and our church/temple communities, poisoning them as surely as any landfill will toxify the area around it. Third, a barrier is created by these contents that prevents our access to the zones beneath them. When this happens, we lose our connective experience to the familial, cultural, and preconscious zones of ourselves. We lose the plants, insects, animals, the planet, the cosmos, and God.

Subzone 2: The Familial Subconscious (Physical Corollary: One's Family)

As Kahil Gibran once said, "Who among us has not cause to weep over their parents?" We are all imperfect. We have all had imperfect parents. We all become more or less imperfect parents if we become parents at all. We all come from imperfect families as well. Families have an enormous, if often subconscious, effect on our personalities. In love a parent might say,

"You're a real operator. Everything you do will turn out OK." Another might say in anger, "You'll never amount to anything." In either case, the parent's comment can have an enormous impact on the child's future. Families shape us at so many levels that it seems impossible to speak of ourselves without speaking of our families.

A good neighborhood is a family of families. A town is a family of neighborhoods. A state is a family of towns. Nations are families of states. The United Nations was conceived as the "Family of Nations." As the living psychic tissue of our families continues to disintegrate, so do our neighborhoods, our towns, our states, our nations, and the world. In the human, primate and mammalian evolution have led to and selected the family as the optimum context for the raising of human young. The family is nest and niche; we shape it and it shapes us. As the primates and mammals become extinct, the human family unravels. I believe that there is a connection here. Primate, mammalian, and even reptilian evolution have provided the architecture of the human family. The family is a biological reality before it is anything else. We can create entities that we call families, but many of these are cognitive or legal constructs, not families. When we misunderstand what a family is, the family disintegrates. When they disintegrate, so do we.

Subzone 3: The Cultural Subconscious (Physical Corollary: One's Culture)

All cultures ultimately spring from holy books or holy experience. We speak of the Christian West or the Hindu or Buddhist East, the Islamic countries, Native Africa, Australia, or America. However far we've drifted from the original sacred context for our society, deep within, the sacred idea and tradition are there, providing a context for our family experience. We cannot speak of ourselves without speaking of our cultural inheritance. Whether our relationship is one of acceptance of our cultural values or one of rejection, the effect of the culture is profound nonetheless. To a large extent, one's culture provides the psychic context for our families, ourselves, our beliefs, our feelings, and our very thoughts. I'm not saying that we are predetermined completely by our cultures or even by our families. Obviously, we can change.

What I am saying is that both leave their impress on us in many ways; some are conscious, but others are not always fully conscious. Like our thoughts, our feelings, our beliefs, our experience, and our families, our cultures can be thought of as a zone of our personalities. Many of our cultural assumptions determine and regulate our relationships to the nonhuman life community, to earth and cosmos. These assumptions are the psychic tissue that either bonds us to the precultural world or, in being torn, alienates us from it. In my experience, much of therapy and even religion ends here. Fundamentalism of any and all kinds denies the values of the cultures of others. Therapy often plays down the importance of culture in one's psychic life, preferring to leave cultural issues to sociology. But we are our cultures! We must explore our cultural assumptions; they will lead us back to a species relationship with earth. Thomas Berry puts it this way: "We must reinvent the human at the species level." When we do so, we begin to hear the music of nature. Wonderfully, we begin to hear the voice of the primates. Shamanism is the door in and out of the cultural subconscious, the door into the garden of the primates and back again into the architecture of culture.

Preconsciousness

Subzone 1: The Primal (Physical Corollary: The Primate Family)

As we wander out through the crumbling arches of the cultural subconscious and onto the grassy savanna of the primal preconscious, we begin to encounter aspects of ourselves that we share with our earlier primate cousins. Because the remaining primates embody not only the physical (e.g., DNA/RNA) history of the protohuman primates as a whole, they embody the psychic tissue created over the millennia of their tenure on this planet. As such, they provide a zone in the human preconscious. They embody, in other words, the psychic conditions that were essential for the human form of consciousness (i.e., self-reflexive consciousness) to emerge. As the various primate groups become extinct or near extinction, the very supports for the human enterprise crumble beneath us. We watch the huge cracks in the cultural architecture widen, never suspecting that the crumbling primate foundations beneath them are the reason. The individual souls of these beautiful, sensitive, and intelligent animals comprise the supports for our own. When we destroy them, we assault our own psyches. No wonder we cannot heal the epidemic of mental illness! As we treat consciousness, as we try to patch the subconscious, we disassemble the preconscious.

The bipolar religious battle between the "creationists" and "evolutionists" comes down to the belief or disbelief that human beings evolved from earlier primates. Personally, I think the biblical evidence is in favor of the evolutionists. In my opinion the third chapter of Genesis can be read in more than one way. If we drop the title "Fall of Man" and add "The Ascent from the Garden of the Preconscious" (there is no name in the Hebrew Bible), chapter 3 delivers a whole new set of meanings. As the human partakes of the tree of knowledge, an ascent is made into the consciousness of nudity. The innocence of the garden is left behind, and the timid beginning of civilization begins. If this is so, then to scorn evolution is to scorn the very handiwork of God. If we are from the primates, then that is the way that God willed it. Humility dictates that we praise God for this wonder and recognize our primates as ancestors in the evolution of the Holy Spirit.

If we are not so distant relatives of the primates, then it is no wonder that we should share so many commonalties with them. These commonalties can be observed in beauty parlors and in barbershops everywhere, in our fear of heights and of falling in our tendernesses, our aggressions, our territoriality, our humor, our physical structure, our bodily functions, and many more ways. As we journey beyond the crumbling architecture of civilization and enter the grasslands and jungles of the primates, we can expect to discover, to uncover, a whole layer of ourselves. The primate family is the psychic architecture of the preconscious, which supports our civilizations, our families, our beliefs, our feelings, and even our thoughts. The primates embody what C. G. Jung calls the archetypes, the very organs of our preconscious minds. As we eliminate the primates globally, we disassemble the foundations of the architecture of our own minds. Is the epidemic of confusion, meaninglessness, violence, and fear we are experiencing everywhere a result of this unconscious collapse? If so, then we shall never solve our problems by treating consciousness. If we are psychologically dependent on the natural world, then we must allow the natural world to heal if we are to be sane.

Subzone 2: The Mammalian (Physical Corollary: The Mammals)

Milk, blubber, rollicking sexuality, penises entering into the warm pink interior of the female—two cells join in the dark wetness, giving rise to galaxies of cells that explode into life within her, tearing her in their birth. Celebrations in the oceans, rituals of ecstasy on the land, laughter and love sonnets. How do we even begin to understand ourselves without reference to the other mammals? The present confusion between the sexes is rooted in the trite attempt to understand and define ourselves in a legal context, with little or no reference to our deeper primate and mammalian nature.

Can we ever hope to heal ourselves culturally or ecologically if we continue to deny truths about our mammalian bodies and psyches? What good is political or economic equality if they are based on a firm understanding of who we really are? They are meaningless abstractions, ideas hung in air. As we go on insisting on definitions of ourselves that are based in the cognitive zone of the psyche only, we shall continue to tear a deep lesion between the upper and lower psyche, between the body and the soul. The result can only be pain and madness.

All species evolve in ways that support the continued existence and emergence of their own kind. We in the West, especially, have so shaped our "civilized" lives that little or no reference has been made to our "pre-civilized" natures. We go off for marriage counseling, stunned at our loss of the ability to perform own own species' mating ritual in ways that will provide a nesting to protect the next generations. Our children stagger around looking for someone to parent them. They feel an emptiness so deep they cannot even name it, so painful they cannot even touch it. Somewhere in them they know that we have already killed their brothers and sisters under the bogus claim of rights that simply are not to be found in our Constitution. They kill themselves, overwhelmed by the emptiness they feel in the deep architecture of their souls. They are raw in these deep recesses of the psyche where the basic needs of cultural mammals are partially or wholly unsatisfied. We can't go on this way! We must stop listening to lies that we know are lies from fools that we know are fools! Our children are dying! Our children are dying! Our children are dying! It's time to stop "keeping it positive"! It simply is not positive! Am I an alarmist? Fine, I'm an alarmist.

Including the physical and psychic history of the primates but larger than and preceding it are the mammals. Following the extinctions of the larger reptiles, the mammals began to diversify, filling the niches left by their giant ancestors. With the mammals came a whole new level of affective consciousness. There is a tenderness and concern in mammals that we see only in rudimentary form in the reptiles and birds. Perhaps the reason for this is that mammalian young are carried within the bodies of their mothers and are fed directly from those same bodies in their early lives. Perhaps these create a bonding that evokes physical caring to which the organisms respond with warmth and gratitude. We see this tenderness all across the spectrum of mammalian experience. The total affective history of mammalian evolution is drawn up into consciousness in the mammalian cortex of the human. This affection is drawn into conscious expression in humans as well as they care for their own young and even for each other.

Humans considered the greatest among us, such as Christ, Buddha, and the other great religious teachers, are those who have broadened this mammalian caring outward to include all humans. Mother Theresa is a modern example of this human/mammal phenomenon. Another, one who includes all mammals, all species, earth, cosmos, and all beings is Thomas Berry.

What then does it mean when we drive mammal species into extinction? It means that this mammalism quality of earth is diminished. It is withdrawn from the primates (primate mental illness is on the rise, especially in zoos) and thus from the human. Why are we seemingly unable to stem the epidemic of family violence? Why are we seemingly losing the ability to parent our young? Why are we losing the glue that bonds us into spiritual tribes? Why do we create bogus rights, supposedly in our constitutions, that allow us to kill our own children before they are born? Why do we consider that a freedom? Because the psychic tissue of affection is frayed in the deeper recesses of our psyches. When we destroy the land and sea mammals, we disassemble our own ability to be human.

Subzone 3: The Biogenetic (Physical Corollary:
Reptiles, Amphibians, Birds, Plants, Early Life Forms)

Frogs are disappearing globally. I'm not speaking primarily of those who die in polluted waterways but even others found in pristine environments. Some scientists believe that their exposure to ultraviolet radiation from the sun due to the loss of ozone in the upper atmosphere has caused problems with their skins. There are several theories. One fact remains whatever theory we adopt. *The frogs are dying!* As we continue our journey into the precultural grasslands and waterways of our psyches, we can see that the amphibians form a crucial link between terrestrial and aquatic evolution. Amphibians embody the psychic tissue that links us to our watery beginnings.

The amphibians link the fish to the reptiles. Our own brain stems link us to the reptiles (more bad news for creationists). We could go so far as to say that our brain stems are functional members of the reptile community, the physical link to premammalian creation. It was a reptile who spoke to Eve. In the brain stem are protected the vital life functions of heartbeat and breathing. From the reptilian core of our brain we receive the news from the whole planetary community. It is from this depth that we are warned of the danger we are creating for ourselves. Here we gather up the wisdom of the whole of biological and prebiological time. The wisdom and skills learned by all these creatures over billions of years of evolution provide the context for mammalian, primate, and human cultural evolution. This wisdom and these skills arrive in consciousness in humans and their cultures through the brain stem, the most primal zone of the human brain psyche. Niche wisdom. habitat wisdom, territoriality, the ability to process sunlight into food and vision—these life forms provide the physical and psychic tissue that bonds the mammals, primates, and humans to the earth herself. When they begin to die out as communities, they are telling us that the web is fraying, that something is desperately wrong.

What do we lose when the reptiles disappear? To return to our discussion about boundaries, could the breakdown of boundary wisdom be the result of the breakdown of territorial wisdom gained from the millennia of reptilian evolution? What do we lose when the birds disappear? Could bird song be the rudimentary root of human melody? Is that why modern music sounds fractured and dissonant? What do we lose when the frogs and other amphibians die globally? (This is happening as I write.) Do we lose the psychic tissue that links us to the early life forms below and the reptiles and mammals above? What about the plants and unicellular life forms? When we lose them, do we begin to lose our ability to connect to earth or sun? Without this link to earth and sun, human life will evaporate. Early life forms are the living tissue that bonds earth life to the sun. When that tissue is torn—and we are tearing it—we will first go mad . . . then we will disappear.

The fish are dying! So many species of fish are already economically extinct. So many others are biologically extinct. They go belly up by the thousands and millions. Again, they are telling us that something is terribly wrong. Like me, the fish are alarmists. They too are archetypes in the human preconscious. They have given us our backbone, which they in turn received from even earlier life forms such as worms. They were and even now are a huge population comprising a very important zone of our minds. As they disappear, a complete zone of biowisdom disappears with them. That zone is a deep layer in the architecture of our souls. As it collapses, the cracks travel upward through the mammalian and primal zones and into culture. We sit here in consciousness watching as our culture collapses into violence and madness. Never would we have expected that dead fish had anything to do with our difficulties. When and only when we allow the fish, the amphibians, the reptiles, to reflower will we watch ourselves and our world become sane again. If culture is embedded in the larger life community—and it is—then how can we expect to avoid stress when the whole community is itself stressed?

Before the fish came the worms and even earlier life forms. Earliest of all came the prokaryotic single-celled life forms. These are the primary zones of the biogenetic. They came forth from the prebiological earth. They are the physical and psychic ligaments that bond us to the planet. They first created the biowisdom of linkage. As we destroy them we destroy our connection to the earth. The blue-green algae are responsible for the photosynthetic bond that exists between earth and sun. All life is mineral, water, and light. If the bond between earth and sun is too severely severed, life will evaporate. Already we see the potentially fatal effects of our intrusion into earth-sun dynamics. That intrusion is felt as a deep pain within us and is expressed as fear of the sun, our life source. Unless we allow the earlier life forms their watery and earthly habitat, this interior psychic pain of disconnection and alienation will continue. The earliest life forms link the biogenetic zone of the human preconscious to the geogenetic zone. We will continue to feel like strangers here until we allow our minds full relinkage to earthmind. The earth does in fact belong to us *and* we belong to the earth.

Subzone 4: The Geogenetic (Physical Corollary: Earth)

Life emerged from the wisdom of earth. Life's foundation and source, life's sustenance, life's plan and ability to propagate is the earth. Still, for at least ten billion years there was no Earth, no Sun or planets as we know them. Our solar system existed only in potential. All throughout those ten billions of years Cosmos was emerging in the countless galactic, stellar, and planetary forms that we call the universe. The geogenetic preconscious is the intelligence, the creativity, the mind out of which Earth emerged and self-organized. If early life is our linkage to the earth, then Earth is our linkage to Cosmos. The mind out of which Earth emerged is yet another, still deeper zone of our mind. We stand stunned and fascinated before the night sky precisely because as we gaze into space we are gazing into the depths of ourselves, of our preconscious mind. Our space programs are as much a psychospiritual journey as are our "religious" journeys. They are two phases of the same journey. In looking "up" we are looking at ourselves and our common source. Out of Cosmos emerges Earth. Out of Earth comes life. Out of life comes culture. We are one not only with the material universe. We are one with the mind of the universe. To eliminate the contact with any zone of our minds is to be diminished. To eliminate or impair the history of life is to impair our own minds. To lose linkage is to lose ourselves. To lose ourselves is to live in unreality, and that is leading us into madness.

Subzone 5: The Cosmogenetic Preconscious (Physical Corollary: Cosmos)

Phase 1: The Explicate (Immanence). We are surrounded by Cosmos. We are embedded in Cosmos. We are Cosmos thinking. Physically, we know that there are billions of galaxies in the Cosmos containing billions of stars. What vast beauty is this—creativity and order at such an order of magnitude that to contemplate it is to be stunned. Our galaxy, Sun, planets, Earth—life and ourselves suspended in the midst of this! Prayer and meditation are the only attitudes appropriate in approaching such mystery. Prayer and meditation, these earliest ways of knowing, are the only final way in which we can approach our own mysteriousness. Still, if we are ever to know ourselves, we must come to know this place in ourselves, this place from which moment by moment we spring "fresh from the Word." As T. S. Eliot put it, "We are here to kneel where prayer has been valid. And prayer is more than an order of words or the conscious occupation of the mind praying." The mind praying. Mind is Logos. Mind is Cosmos. Mind prays in us.

Phase 2: The Implicate (Transcendence). Finally, the Source: God. As we come to know ourselves in the cosmogenetic zone of the preconscious, we come to know God. As we come to know God, we come to know ourselves. We might call God Krishna, Wakan Tanka, Yahweh, Mind, Holy Spirit, Tao, or whatever else. Source is God: God is Source. If we are to know the source of ourselves, we come to know God. In that flash of awakening that is our memory of the holy fires of Genesis stored in our depth, we come to know our own holy beginning. But because we can know Source in every moment of time, in every inch of space, we are able to travel back behind the fireball event to our paradisal Source. In that awakening it becomes evident that that transcendent Source is the foundation of immanent Cosmos, that they are two phases, one eternal and one temporal of the same living, personal, loving, communicating God, God Most High, God Most Present. In Cosmos, God is clothed in matter. In all humanity, God thinks about God. The great monotheisms of the earth are correct: there is only one God. The many polytheisms of the world are also correct; there are as many gods as there are beings in cosmos.

I am saying that every human psyche is the whole cosmos and that we all preexist cosmos in Paradise. In the Buddhism of Nichiren Daishonin as taught by Daisaku Ikeda, this realization is called Nam Myoho Renge Kyo. The words themselves are holy ground. It holds that there is no hard-and-fast boundary wherein we can say that anything begins or ends. Every thing is everything! We must come to understand this if we are ever to create a psychology adequate to our present needs. We need a psychology that includes client, therapist, family, culture, the animals and plants, earth, cosmos, and God. This type of counseling is not without precedent either. When shamans heal, earth and heaven are called into the experience. If Christianity were led by true spiritual authorities, not ordained functionaries, this type of healing could begin today. How would we outline a counseling theory for today?

To practice sacred process ecopsychology, counselors must be three things:

1. They must be adept at affective therapy.

2. They must be spiritually adept in at least one tradition and respectful toward all others. Lack of respect for other traditions signifies a shallow understanding of one's own.

3. They must be ecological activists.

I'm not convinced that counseling techniques can really be taught. Some people are able to see into the souls of others; others cannot. I've not personally met many who can. Those I have met can and have benefited from the study of counseling theory because counseling theorists are people who do possess the gift and document their findings in their theories. Certainly, we can all benefit from familiarity with their experience. Most people will benefit from counseling only if the counselor is truly open at the affective level. This openness is experienced in the solar plexus area of the body as a pain or resonance associated with the joining of two souls. Without that connection at the sternum zone of the body, people get stuck in cognitive prattle: analysis paralysis. I've seen people who were in analysis for a decade and emerged worse for the experience and the expense. One cognitive technician preventing another from entry into the depth of the soul will not heal that person of much at all. Affective skills will lead to deeper zones of the soul where much pain is stored in the various layers in the soul's architecture. When these are cleared, the Source emerges in the person, and an awakening to truth is experienced. Truth is health. Truth is vitality.

Traditional Western religions are dying it seems. There are many reasons. The most important one is, as I have indicated, that religions are often led by the spiritually shallow. The shallow are threatened by the depth of those who have true spiritual experience. When one has spiritual authority, one cannot be manipulated or controlled. The shallow become threatened by their freedom and persecute them. The very best are thus often marginalized by the least important of religious functionaries. Fundamentalisms are the diseased leftovers when religions meant to be prophetic and revelatory become tepid and boring. As the Tao Te Ching puts it, they then seek to enforce themselves with rolled-up sleeves. These fundamentalisms are the diseased right pole of religious bipolarity. Most of what we've come to call New Age is the disease of the left pole. Heretical forms of Western and Eastern as well as native religions, self-centered egoisms inflating rather than desiccating the self, yapping uniqueness wherein everyone is so very special that one can agree with no one else about anything, exclusivity wrapped in inclusive language, the trite rejection of traditional religious systems that have given birth to and have guided whole civilizations over millennia are some earmarks of the left. *Option Three* is our only hope.

Option Three is an authentic religious renewal from within, from the bottom *and* from the top. If we do not understand our traditions, we shall never come to understand our institutions: the latter are born of the former. Sony cannot guide Japan; a revivified Buddhism can. This renewal of religious sensibility must include all that we now know about cosmology, geology, biology, anthropology, comparative religions, psychology, and Alcoholics Anonymous. Alcoholics Anonymous is a successful experiment in human love. It has proved that people from every background, every continent, and every sexual persuasion, people who might not like each other, who might even hate each other, can form a fellowship free of exclusivity and other forms of violence focused only on health, service, and experienced relationship with God. Alcoholics Anonymous is the most significant spiritual movement in the world today. Option Three must contain a deep and authentic understanding of the feminine at all levels of reality. We are all worn out by what Eliot called "the absolute paternal care that will not leave us but prevents us everywhere." This does not mean, however, that we must support what Mary Rosera Joyce has called "women's special form of killing." Nor must we support the "right" of men to hire assassins from the medical establishment to dispose of their inconvenient children. The so-called legal right to kill preborn humans is a lie. As Martin Luther King Jr. put it, "a lie cannot live forever."

But what about the preconscious zone of the psyche? How does the therapist function there? First, the client must know that the counselor is an ecological activist. What good is it to deal with the cognitive, affective, and creedal zones? Why do family or religious counseling if we show no concern for the foundations of the psyche rooted in the created order? Can I pretend to be concerned for your life if I have no genuine concern for the context of your life or for your future? All the "sensitive" pretense will have no lasting effect if by my life-style I am advocating for the destruction of my client's deepest psyche. We must be trained counselors, spiritual adepts, *and* ecological activists, defending the deep psychic roots of those who come to us seeking care of the soul. As client and counselor alike rediscover the source of our lives, as we rediscover the animals, plants and earth, as we rediscover our capacity for love, we will rediscover a love of all of life. Wonder will well up in us as we awaken to our true dimensions as sacred beings. As the creation experiences this renewal of our love and concern for the least of God's beings, the architecture of our own souls will rejuvenate. We will have returned to our humanity, to sanity, and to truth.

So finally, we come to our question earlier as to the name of this disease that is destroying not only the cultural world but the natural world as well. Every generation has its courage probed by a major challenge to its health and well-being. The confrontations with fascism and communism are two examples. The confrontation is the perennial struggle between good and evil. The name of the disease we face is *evil.*

The Universe Story

Its Religious Significance

I write these words in the hill country of the northern Appalachians, at the eastern edge of the North American continent, some miles inland from the North Atlantic Ocean, during what might be considered the terminal phase of the Cenozoic period in the geobiological history of the earth. To get some understanding of the nature and order of magnitude of what is happening just now, it might be helpful to situate our thinking within this Cenozoic period, which can be dated roughly as the past sixty-five million years of earth history.

This is the lyric period in the entire history of the planet. During this period the flowers and birds and forests and all the mammals, such as we know them, came into being. As this continent drifts westward away from Africa and the great Eurasian continent, we are opening the North Atlantic Ocean ever wider as we move across the Pacific Ocean and approach the Eurasian continent on its eastward borders.

We need to think of these things, of where we are and what is happening on this larger scale as well as on the smaller scale of the territory we occupy and the times in which we live. We need to think of the northern Appalachian Mountains that surround us. We need to think of hardwood forests and the magnificent white pines that once covered this region, the region with perhaps the greatest display of deciduous trees on the planet. The more we think of these things, the more we are caught up in the wonder and mystery of the world about us, the more evident it is that we live amid a vast celebratory process, a kind of colorful pageant beyond anything that we as humans could ever have imagined.

The more we consider all this, the more evident it is that the universe throughout its vast extent in space and its long sequence of transformations in time can be seen as a single multiform celebratory event. Our human role, it would appear, is to be that being in whom the universe reflects on and celebrates itself and its numinous origins in a special mode of conscious self-awareness.

While this pertains primarily to the universe itself, it pertains in a special manner to our experience of the planet Earth, one of the nine planets in our solar system. All of these were originally the same, composed of exactly the same material, yet only Earth came to express itself in such color and form and life and movement, in such taste and fragrance as we observe about us. Only Earth of all the planets in our solar system was able to burst forth into such magnificence.

Mars turned to rock because the gravitational pressures could not produce the inner heat required to create the turbulence needed for air and water and life development. Jupiter remained the turbulent fiery mass of gases that it was in the beginning. Its gravitational pressure was such that nothing firm could take shape. Jupiter has no surface such as we find on Earth. Of all the planets only Earth had the proper balance between turbulence and restraint that enabled the planet to bring forth the amino acids, the living cells, all the superb organisms that inhabit the earth, and finally ourselves.

Earth, too, was just the proper distance from the Moon so that the tides could keep the seas in motion. Otherwise, if the Moon were closer to Earth, the tides would overwhelm the continents. If the Moon were somewhat more distant, there would be no tides, the seas would be stagnant, and life could not come. So in relation to the Sun, Earth is situated so that the appropriate differences of temperature could exist between the arctic and the tropics, and the vast variety of other climatic conditions could exist to shape the diversity of life as we find it.

There is something wild and unfathomable throughout this entire process, something that evokes awe and wonder at the source from which all this came into being, something that invites us to participate in this vast celebratory process. From earliest times humans in the temperate parts of the world have sought to enter into the ever renewing cycle of the seasons through ritual celebration of the springtime renewal, when new life appears throughout the plant and animal kingdoms. Spring is especially significant for the mammals whose cycle of gestation has taken place throughout the winter months and who are ready to bring forth their young. Spring is the time that mating rituals take place, especially the gorgeous rituals of the birds.

These human celebrations became ever more sophisticated as the great urban civilizations developed and the ceremonies were elaborated. I mention all this as integral with the Cenozoic period, for it seems that the human could come into being only at a period when the planet was at such a gorgeous moment in its expression of itself and also at a moment when the human could enter into the larger functioning of the universe through some form of ritual celebration coordinated with the great liturgy of the universe itself.

It would seem that the coming of the human mode of consciousness could have occurred only when a world with the brilliance of the advanced Cenozoic period had set the stage. Perhaps this was needed to awaken human intelligence, imagination, sensitivity. Yet if this brilliance was needed to excite wonder, it was needed also as a healing for the sorrow of life that would inevitably result from the burden of the human form of intelligence and freedom of choice in actions. The human had to shape itself to a degree far beyond that of other modes of being.

While other modes of life are guided in their self-expression through their genetic coding, with relatively little further teaching or acculturation, there is a further self-formation of the human, a cultural coding mandated by our genetic coding. This required a special mode of self-invention on the part of the human. Once the human was brought into being, there was a several-million-year period when the earliest forms of intelligence were elaborated in the shaping of implements, in the discovery of fire-making, in the shaping of social order, in learning the arts of dealing with the nonhuman world, especially in dealing with all the spirit powers perceived throughout the surrounding world. It was a world of person presences. Every being was to be addressed as a "thou" rather than an "it." Finally, there was the invention of spoken language, which begins we know not when. We know only that this greatest of human inventions most likely occurred within the past hundred thousand years, perhaps as recently as the past fifty thousand years.

We need not go through the long narrative of the period leading up to the Neolithic village life of some twelve thousand years ago nor to the shaping of the great urban literate civilizations of the past five thousand years. These civilizations continued in the pattern of earlier human development. The discovery of writing was decisive in its consequences, for the various traditions fixed their revealed texts in written language that then came to control the greater part of the human venture over these past several thousand years.

In the past few centuries, however, since the time of Copernicus, a new area of consciousness has awakened in Western intelligence. We began to look more intently at the universe in its material structure, both the forces that swing the starry heavens through their orbit and the structure and functioning of forces that govern life on Earth. Suddenly, we discovered that the heavens did not move in circular but in elliptical order, that neither the heavens nor the earth were formed of eternal matter, that the smallest particles composing the material world contained immense quantities of energy. But above all, we discovered that the universe came into being through a vast period of time, through a sequence of transformations leading from the simpler to the more complex, from lesser to greater manifestation of consciousness, from lesser to greater freedom of action.

These discoveries gave to the human a range of power over the functioning of earth such as was never dreamed of in former times. We discovered that we could use earth for our own indulgent purposes. Soon we turned the entire human venture into an assault on those planetary processes that have over the millennia brought about all these wonders that we have outlined here. It is such a poignant moment as we look about us and observe that we are terminating this period of immense creativity, the Cenozoic era, the period that has for sixty-five million years brought about the grandeur of the world about us. We generally talk about our times in more limited terms, saying that we are at the end of the Enlightenment period or at the end of the medieval period or terminating Western civilization. We even think at times of the ending of the human mode of being due to the degradation of life conditions.

Yet we must consider that what is happening now is not some change such as occurred in the transition from the classical Mediterranean phase of Western civilization to the medieval period or from the medieval to the commercial-industrial civilization of the past two hundred years. Nor is the present situation something that pertains to the survival or destiny of simply the human. We are apparently at the end of a geobiological period. We are altering the chemistry of the planet, the biosystems of the planet, even the geological structure of the planet, all in a deletrious manner.

For the most part this has happened in my generation. When I was born in 1914, the planet and the North American continent were severely damaged but perhaps in a manner that could have been recovered from to an extensive degree. Now the planet has been damaged far beyond what occurred in the earlier part of the century, so damaged that the children of the immediate future will live amid the ruined infrastructures of the industrial world and amid the ruins of the natural world itself.

When I ask myself how to explain what has happened, I can only answer that my generation has been austitic. My generation had no effective capacity for communicating with the nonhuman world. Earth was seen as an inexhaustible resource of materials for human use and consumption. The nonhuman world had neither honor nor rights nor any sacred mode of its being.

Here in America we were heirs to the English tradition of jurisprudence, which is deeply concerned with life, liberty, and pursuit of happiness for humans at the expense of the natural world. Humans were protected in their liberty to own and exploit property for whatever purposes they wished. Yet mountains had no rights to their grandeur, rivers had no rights to remain free of pollution, the salmon had no rights to their spawning places, birds had no rights to their habitat nor to protected access along their migratory paths.

The rights we enjoy are determined by the Constitution. Yet nothing in the Constitution or in the Bill of Rights recognizes any rights possessed by natural modes of being. The National Geographic Survey was instituted early in the history of the nation, not for the purpose of a more profound communion with the natural wonders of the North American continent but for the purpose of distributing the land to private ownership, with no inherent responsibilities other than not to infringe on the rights of other humans dwelling in the same region. That land should be owned in accord with the nature of the land and the integral mode of its functioning was never a question.

Exploitation was the preordained way for humans to bear themselves toward the surrounding world: the continent must in some sense be re-engineered and its power appropriated; otherwise it was simply wasted. Not to dam the western rivers—the Colorado, the Columbia, the Snake, or the river that flowed through the Hetch Hetchy valley of California— was wasteful. Not to exploit the Tennessee with a long series of dams was to refuse the power and the water offered there. Not to force the soil with fertilizer was to deny ourselves an increased harvest. Not to pave the roads was neglect. Not to take the petroleum from the earth was to reject a God-given opportunity for bettering human life, despite the fact that nature had stored the carbon in the petroleum and in the forests so that the chemical constitution of the air and the water and the soil could be worked out in some effective manner. That humans had rights to do what they pleased was self-evident, not to be contested.

To explain such autism it is not sufficient simply to go back to nineteenth-century industrialization or to Newtonian physics or even to Francis Bacon or Descartes. The origin of such autism requires a more profound explanation that would push our inquiry back into the anthropocentrism of the Hellenic world; back also to the biblical world and the scriptural foundations of our Western life formation; back to the two great commandments, love of God and love of neighbor; the fulfillment of the Law and the Prophets with no reference of any relation with the world about us.

That our religious traditions, our humanistic traditions, our educational programs, our jurisprudence, and the other shaping forces of our society all contributed equally to this autism might be too heavy a position to propose, but to note that none of these traditions was able to prevent this autism and the destruction emanating from within our Western civilization seems entirely appropriate. To say that all of these traditions were excessively committed to anthropocentrism also seems a proper conclusion. To say that they all favored processes that led to the present disastrous situation may even be defensible.

Certainly, none of these traditions has protested the devastation in any comprehensive manner, nor has it altered its basic orientation in any substantial manner. That is the difficult side of our present situation. There seems to be a stand-off attitude, an attitude of noninvolvement, even what might be considered a pervasive denial of the real magnitude of the difficulty.

Here we might consider just where we go from here. I propose that we need to go from the terminal Cenozoic to an emerging "Ecozoic" period, defined as the period when humans would be present to the earth in a mutually enhancing manner. This is the clearest way that I can express my own sense of the possibilities that are before us.

I prefer the term *ecozoic* to *ecological* since this enables us to place the coming geobiological period in its proper context in the sequence from the Paleozoic (from 600 to 220 million years ago), to the Mesozoic (from 220 to 65 million years ago), to the Cenozoic (the past 65 million years), and what now might be termed the Ecozoic period. This might now be accepted as the proper sequence in articulating the ages of the earth.

The term *ecozoic* is appropriate because it indicates that we are concerned with life forms themselves, not simply with our understanding of the life forms. But most of all, the term *ecozoic* gives some feel for the order of magnitude of what we are about. Our greatest failure at present would be to underestimate the exact magnitude of the issues that are before us.

What I miss most at present is the realism needed in evaluating the nature and order of severity in the challenge we face. Despite all the renewal of the Earth Movements, our society generally is still, it seems, in a period of denial. What is before us is too overwhelming. We are still reacting in a manner similar to that of the autistic child, who because of some psychic trauma has closed itself off from communication with the outer world. So we have closed ourselves off from any intimate feeling rapport with or even any rational understanding of our relation to the outer world. Indeed, if we were not so closed off, if we had some rapport with other modes of being, we would not be able to do what we are doing.

Yet we are beginning to listen when the soil tells us that it will grow our food if only we will assist it in functioning according to its own rhythms and in accord with its own needs. So too we begin to understand the infinite abundance of marine life in the seas that could feed us forever if only we permit the abundant marine life to multiply in accord with its inherent nature. We are finally learning when the natural world tells us that it cannot fulfill its role in sustaining us if we interfere with its proper modes of functioning.

We would expect the universities and the religious establishments to have guided us long ago, for these carry the humanistic and the religious wisdom of our traditions in their most exalted form. Yet even when biologists such as E. O. Wilson indicate that the extinctions of life occurring now have not been experienced on the planet since the termination of the Mesozoic era, even with such validation of the severity of what is happening, the universities have shown almost no willingness to shape their academic and professional programs to meet such a situation, thus leaving the students in a certain ignorance of the real-life context in which they must function in their professional lives. As with the professions generally, the move from the terminal Cenozoic to the ecozoic mode of functioning is a transformation beyond their capacities for adaptation, even though it is increasingly clear each day that the present modes of functioning of all the professions are leading us ever deeper into a tragic impasse.

But while we have used our modern knowledge in destructive ways, we find that the knowledge itself is valid and has given us a new story of the universe. This is the supreme achievement of modern intelligence, even though it has given us this knowledge as secular, materialistic, without inherent meaning. Yet once we realize that the universe has had a psychic-spiritual as well as a physical-material aspect from the beginning, once we realize that the human story is inseparable from the universe story, then we can see that this story of the universe is in a special manner our sacred story, the story that reveals the divine, the story that illumines every aspect of our religious and spiritual lives as well as our economic and imaginative lives.

Once we recognize this mystical dimension of the universe, we can appreciate the unity of every being out of this same primordial origin; we can see that every being in the universe is cousin to every other being in the universe. This is especially true of living beings who are descended through the same life process. Whatever the causes of our present situation, our need now is for a program that would enable us to manage the transition into the Ecozoic period in an effective manner. There are a number of conditions that must be fulfilled if we are to make this transition to a period in which humans would be present to the earth in a mutually enhancing manner.

First, we must understand that the universe is a communion of subjects, not a collection of objects. This implies that we recover our primordial intimacy with the entire natural world. We belong here. Our home is here. The excitement and fulfillment of our lives is here. However we think of eternity, it can only be another aspect of the present. The urgency of this psychic identity with the larger universe about us can hardly be exaggerated. We are fulfilled in our communion with the larger community to which we belong. It is our role to articulate a dimension of the universe. In a corresponding manner our smaller individual self is fulfilled in our larger self, in our family self, our community self, our earth self, our universe self. That is why we are drawn so powerfully to inquire into and understand and appreciate the stars in the heavens and the wonders of the earth. Every being is needed, for every being shares in the great community of existence. The comprehensive community is the supreme value in the phenomenal order.

Nothing substantial can be done until we withdraw from our attitude that every other mode of being attains its identity and value simply in being used by the human. Every being has its identity, its honor, and its value through its role in the universe. The universe is the normative reference, for as Saint Thomas tells us in his *Summa Theologica*, part 1, question 47, article 1, "the entire universe of beings participates in and manifests the divine more than any single being whatsoever." Within the larger universe the planet Earth constitutes a single integral community of existence. It lives or dies, is honored or degraded, as a single interrelated reality. As regards the future, it can be said quite simply that the human community and the natural world will go into the future as a single sacred community or we will both experience disaster on the way.

That the human is a subsystem of the earth system is most evident in the economic order. To advance the human economy by subverting the earth economy is an obvious absurdity, and yet our entire commercial-industrial system of the present is based on this absurdity. Only now is a new consciousness emerging in the economic institutions of our society.

Most difficult is jurisprudence. There already exists in the natural world a governance of the earth, a governance too subtle for us to understand. This governance enables the earth to bring forth the immense variety of its living forms that interact so intimately and so extensively with each other that the well-being of each is fulfilled in the well-being of the whole. This governance has capacities far beyond anything that humans are capable of. Yet this must remain the context into which we assert our human governance. Our human governance needs to function within the context of earth governance, just as our economic functioning needs to be an extension of the earth economy.

As regards healing, we begin to appreciate that the earth is a self-healing community just as it is a self-sustaining community and a self-governing community. It becomes clearer each day that there can be little hope for human healing except through the assistance of an integral natural world. When the earth becomes toxic, humans become toxic. We lose the only context in which we can hope for that vigorous mode of well-being that should be ours.

The next condition for our entering effectively into the Ecozoic era is that we accept our new story of the universe as our sacred story, with a special role to fulfill in this transition moment from the terminal Cenozoic to the emerging Ecozoic. This story enhances rather than negates the other stores of the universe that have over the years guided the course of human affairs among the indigenous peoples of the world as well as in the classical civilizations that have presided over the greater volume of human expression over the centuries. We are, however, at a time when these earlier traditions can no longer, out of their own resources, provide adequate guidance in the task that is before us.

Assuredly, we cannot do without the guidance of these traditions of the past. They provide understanding and guidance that is not available from this story of the universe that we are presenting here. Yet we are faced with vast realms of knowledge and power that require the new range of understanding available to us from this new insight into the structure and functioning of the world that surround us.

We need the story and the dream. We need the story to understand where we are in the unfolding reality of the universe. But we also need the dream, for the dream drives the action; the dream creates the future. The Ecozoic era must first be dreamed. Through the dream comes the guidance, the energy, and the endurance that we will need. For the transition that is before us will cost an immense effort and a wisdom beyond anything that we have known before. Our greatest encouragement just now is that we have begun to dream the ecozoic dream.

Here it is necessary to note that our planet will never again in the future function in the manner that it has functioned in the past. Until the present the magnificence splashed throughout the vast realms of space, the songs that resonate throughout the earth, the luxuriance of the tropical rain forests, the movement of the great blue whale through the sea, the autumn colors of the eastern woodlands—all this and so much else came into being entirely apart from any human design or deed. We did not even exist when all this came to be. But now, in the foreseeable future, almost nothing will happen that we will not be involved in. We cannot make a blade of grass, but there is liable not to *be* a blade of grass unless we accept it, protect it, and foster it. Even the wildnerness must now be protected by us. On occasion the wild animals need our care. So too there is an infinite amount of healing that must take place throughout the planet, healing that will at times require assistance from us; although, for the most part, the natural world will bring about its own healing if only we will permit it to function within the dynamism of its own genius.

Just now we are a transition moment such as neither we nor any other present living being, nor the earth itself, has ever experienced before. We are in the early hours of dusk as night settles over the land. Day is over, and night moves quietly over the land with its healing power. Dawn, when the eastern sky reveals itself in its faint purple glow, night when the sun sinks over the horizon and reflects its light in a colorful spectrum on the clouds of the western sky—these are the sacred moments of the day, the mystical moments, the moments of transformation.

So with sacred moments generally. Our moments of grace are our moments of transformation. In human life our greatest transformation moment is the moment of our birth, a sacred moment indeed, followed through the years with the sacred moment of adulthood and marriage and then the moment of death, itself such an awesome, such a sacred moment.

But if such moments of transformation—in the day and night, in autumn and springtime, in birth and death—are sacred moments, we must believe also that those vast cosmological transformation moments that enable the universe, Earth, and all its living beings to come into existence are sacred moments. Such a moment we observe in the sacrificial collapse of that first-generation star, which formed, in the intense heat of its collapse, the ninety-some elements that were needed for the formation of our solar system and especially the planet Earth, elements needed for the emergence of life. These elements were formed and then scattered with infinite abandon out into space, to be gathered and shaped into our Sun and its nine planets.

This was, I suggest, a cosmological moment of grace, for only through this event did Earth or life or the human form of consciousness become possible. So too, there are other cosmological moments of grace, such as the moment when photosynthesis was invented. These moments are indeed moments of grace, moments of divine manifestation, on a scale that we seldom think about. Such I would propose is the moment in which we find ourselves just now at the terminal phase of the Cenozoic era. We are in a transition moment, a transformation moment on an immense scale, as we experience the decline of the Cenozoic era and begin to shape some idea of what the emerging Ecozoic era might be. This, we must believe, is a moment of grace on a scale we have never thought of previously.

We begin to recognize in a dim and distant manner what is before us. Yet before my generation moves from the scene and the new generations take over their role in the immense drama of these centuries, we might communicate to them at least that encouragement and that faint glimmer of wisdom that we have attained in these recent decades and provide them with some assurance that the task before them is not simply their task. It is the task of the entire earth community, for only this community is capable of the transformation that is needed.

Finally, I would note that this transition to the Ecozoic era is the great work that we are about at present and for the immediate future. It would appear that each age has some great work that provides for the age its life purpose. This great work enables societies and civilizations to endure the agonies inherent in fulfilling any significant role in the larger historical process. We have, it seems, an immediate, particular work or profession whereby we fulfill our role in the social order, obtain our living, and support our families. Within this context we carry on the great work to be done in the larger historical order; inventing the Neolithic period, building a civilization, establishing a religious tradition, founding a nation, building the medieval cathedrals—these are among the great works of the past.

In more recent times the great work of the scientists has been to discover the large-scale as well as the small-scale structure and functioning of the universe and of the planet Earth. The pathos of our times is that the commercial-industrial establishment of this century thought of itself also as doing a good and great and noble work. Even when it was devastating the planet and bringing the Cenozoic period down into ruins, it thought that it was introducing the human community into a millennial age. Such is the deep cultural pathology that we are called upon to heal. Such a poignant moment. We already had a glorious world; but we did not recognize it or know how to relate to it.

History has chosen us to begin the great work of the twenty first century, to initiate the Ecozoic era as a remedy for the cultural pathology of the present and to initiate the period when humans would be present to the earth in a mutually enhancing manner.

Afterword

A Journey of the Heart

There is an old Hassidic tale of a teacher who offered this challenge to his students: "I will give one gold coin to anyone who can show me where God is." After a long silence a young student spoke up: "I will give two gold coins to anyone who can show me where God is not."

For many of us living in this, the final decade of the twentieth century, caught somewhere between the polarities of materialism and spiritual values, the world may appear to be either totally devoid of any meaningful understanding of God or simply bursting forth with the presence of the sacred. Perhaps most of us vacillate somewhere between those extremes. On our best days, when we've had plenty of sleep, the weather is inviting, and we have sufficient time to simply "be," we are keenly aware of the utter splendor that surrounds us in plant and animal, starry night and crashing wave, laughter of child and embrace of lover. But in the rush and press of life that sense of wonder and deep awareness of mystery may elude us.

I am reminded of a warm spring day in New York City, where I served my first pastorate fresh out of seminary. The pace of the city and the demands of my work often tested my theology and stretched my capacity to feel connected to God's good earth, its seasons, cycles, and miraculous ability to heal the weary spirit. On that particular day, feeling in need of rejuvenation, I wandered into Riverside Park to sit and gaze at the Hudson River and try to regain my perspective. Moving off the noisy street, I found a quiet bench. Only a few moments later, while my mind was still preoccupied by the tasks that awaited my return, I noticed a disheveled-looking man moving along the sidewalk not far from me. His dirty clothing, incoherent ramblings, and characteristic plastic bag quickly identified him as one of the city's many homeless who spent their days gathering discarded cans and bottles to redeem at day's end. After diverting my eyes for a time, my gaze eventually wandered back to this "lost soul" just as he stooped to examine an object on the ground not twenty feet from me. I then noticed that it was not merely another aluminum can his hand now cradled but a young bird, apparently having fallen from its nest in the branches above. To my utter amazement the man muttered a few words, apparently to the tiny creature, then tenderly lifted it back into the tree, returned to his "work" and disappeared off into the park. Later that afternoon I shared the experience with a colleague, who responded that it was clearly the erratic behavior of a "lunatic," discounting my suggestion that what I had witnessed was a human being living in the midst of the holy and aware of it. To this day that nameless stranger remains for me a symbol of something many of us have lost.

For those who are truly awake to the mystery and "jaw-dropping" awesomeness of our vast universe, fifteen to eighteen billion years in the making, there is the risk of being taken for a lunatic. Those who are made breathless by the intense beauty experienced through dazzling sunsets, the intricacy of a spider's web, or the perpetual, rhythmic dance of the moon, tides, and seasons may find themselves viewed with some suspicion. At least this is true in Western culture, where we have inherited a legacy in which mastery has replaced mystery, consumption of finite resources and unlimited growth have become "rights," and scientific knowing has replaced wisdom drawn from nature and human intuition. This dominant cultural paradigm has distanced us from the very source of our existence, the earth, and has replaced a sense of connection to the physical world with a desire to control the "chaotic" forces of nature. The resultant dualities, experienced by most of us on a daily basis, preclude a sense of reverence for the natural world and undermine our awareness of the interconnectedness of all life. But oh, how refreshing and life-giving are those "lunatics" among us who continue to remind us of the sheer wonder of it all and elicit within us our own passionate, jubilant response.

With the planet experiencing environmental degradation on an unprecedented scale and ecological disaster looming with life-threatening force, the human community is receiving a dramatic wake-up call. This blue and green jewel we call home can no longer accept our shortsighted and self-indulgent choices. In response there is a reawakening in process, as a new, yet ancient story begins to be told. It is earth's story; it is the story of billions of far-flung stars and of tiny sea anemones; it is a story told by mystics and mountains, prophets and planets, scientists and systems. This story speaks of an intimately connected and intrinsically relational universe unlike anything we have dreamed of before. Like sleeping giants we move slowly and clumsily into wakefulness until, with the help of modern-day prophets like Thomas Berry and others, we wipe the sleep from our eyes to witness a brilliant dawn and discover that the "story of the universe is in a special manner our sacred story, the story that reveals the divine, . . . and illumines every aspect of our lives." Even in what seems like the close of a perilous age, there is a global reawakening occurring as countless members of the human community reclaim their connection to the earth and begin to give voice to this sacred and life-affirming experience.

Throughout the preceding essays the richness of this story has been illuminated, the diversity of voice and symbol celebrated. Unlike some stories, those hearing this one are invited to join, not as audience but as participant. At first, it is not an easy story to hear, for one encounters a tale of almost nightmarish quality—of planetary devastation, the loss of species, the unfathomable consequences of overpopulation and ongoing ecological mayhem. Timothy Weiskel is no mere doom-sayer when he calls upon us to realize that we are "in the midst of a global 'extinction event' which equals or exceeds in scale those catastrophic episodes in the geological record that marked the extinction of the dinosaurs and numerous other species." The horrors seem almost too overwhelming, and one is tempted to deny these realities rather than confront them. My work as a university chaplain gives me daily opportunity to witness the denial taking place within young adults who see a world of shrinking possibilities and insurmountable obstacles. If the bomb doesn't finish us off, the reasoning goes, surely ecological disaster will. And it is not only the young who feel the numbing effects of a world perceived to be on the verge of collapse. Perhaps, as Albert LaChance and others suggest, rampant substance abuse and unchecked consumerism are merely the hallmarks of a society dominated by fear and living without hope, a society no longer in sync with the natural world, blind to the sheer wonder of life itself, without a meaningful way to speak of the holy or sacred.

If the story ended here, there might truly be cause for despair. But echoing through these pages one hears of a new age that is aborning, an "ecozoic" age of spiritual and ecological renewal. The voices, coming from varying religious traditions, call us to move from a paradigm of domination, fear, and alienation to a paradigm of partnership, mutuality, and reverence for all living things, wherein spiritual values are reclaimed and divine immanence is reaffirmed. Aware of the emerging, dynamic story of the origins of the universe, these modern-day prophets cry out with compelling challenge to rediscover our rightful place within the natural world, to awaken to mystery and our own deep longing for interdependence. Their voices join in one harmonious chorus, inviting us to humble ourselves, not before altars made by human hands but before all that is holy and beautiful around and within us.

For the theologians, academicians, and spiritual leaders whose voices emerge from the preceding pages, there is little confidence that government, business, or even religious institutions will be able to mandate the enormity of change required as we seek healing for our earth and for ourselves. Instead, they describe a journey that requires a true shift in the way we view Earth and ourselves, how we speak of mystery, and how we choose to live together with the amazing diversity of animate and inanimate life with which we share the planet. Perhaps most fundamentally, it is a journey that involves accessing the deep reservoirs of love and caring within the human heart, trusting our innate awareness of the intrinsic worth of all life. Stephen Rockefeller expresses the belief that so many others share when he writes that "if human beings search deeply enough, in their hearts they know the sacredness of life and the goodness of creation." For too long we have devalued this "heart knowledge" and subsequently are in need of reconnecting with its power to shape a sustainable and hope-filled future. Whether speaking of the Buddhist understanding of compassion and loving kindness, the Abnaki vision of loving all of creation for the miracle that it is, or the biblical commitment to cherish and care for God's good earth, there is the consistent recognition that the hope for our future lies within our ability to listen to our heart's wisdom and "harness the energies of love."

For many of us, this journey of the heart begins with it a new level of mindfulness, simple awareness of our own state of being and of the state of being of those things around us. In slowing down and rediscovering a world of flora and fauna, mountain and swallow, the majestic and minuscule, and the amazing cycles of birth, death, and rebirth, our own capacity for knowing and loving will enlarge in joyful and life-giving ways. Then we will come home to the "green grace" that Jay McDaniel describes as "the healing and wholeness that we find when we enjoy rich relations with plants, animals, and the earth." The mindful life may very well be the starting place for the transformed life.

Recently, a friend told me of receiving a beautiful Christmas cactus as a gift. She dutifully placed it near the window, watered it with regularity, went on about her busy life, and was delighted that it continued blooming month after month with little or no care. It was not until some months later a guest discovered, to the great surprise and chagrin of my friend, that the plant was artificial, albeit a "good" imitation! In her own preoccupation she had missed the truth about the plant and in so doing discovered a new truth about herself.

In a world of innumerable preoccupations that tend to keep us disconnected from the miraculous reality of our surroundings, religious and spiritual traditions offer a glimpse of what this journey may require. These traditions offer us language and symbols that speak of mystery, transcendence, immanence, and love. Without these to guide us on the way, it is doubtful that we will be able to make the kind of changes necessary to awaken within us a new reverence for life and to alter the current course of environmental destruction we are currently racing down. Our disconnection from the very source of our lives, the earth, has resulted not only in ecological degradation but in spiritual degradation as well. In retelling his experience at a workshop on religion and ecology in the Hill Country of Texas, Jay McDaniel vividly illustrates the way nature so often serves as the backdrop for human activity rather than the very context out of which to live and think, work and dream. What he and others suggest is that this alienation from our earth community has dramatic consequences, not only for the planet but also for the human capacity for wholeness and a sense of belonging. The same mentality that produces toxic waste and acid rain also strips the human heart of inner wisdom and ultimately makes enduring peace unattainable. Failing to experience Earth as a living text, be that a Texas vista or a simple houseplant, we forfeit the sheer joy and healing that comes from living in a mutually enhancing relationship. As we begin to experience ourselves as part of the fabric of nature, inseparable and interdependent, we may begin to grasp what TwoBears describes as a fundamental to Native American spiritual life. Our own hearts may then be open to perceive the sacred and in so doing touch upon our own deep capacity for caring.

Reclaiming this deep wisdom may be our greatest challenge as we pursue transformation. In a very real sense this process does not involve the acquisition of more knowledge through traditional scientific means but may require only that we quiet our minds long enough to listen to the silent memory of the very cells of our bodies. For if science is correct in describing the origins of the universe, we may harbor within us a memory forged in a fireball furnace billions of years ago, which speaks of radical interdependence, the glory of uniqueness, and the necessity of diversity. The challenge then becomes one of finding ways to access that memory and bring it to the foreground of our being that we might honor the truly miraculous process that has sprinkled the universe with such abundant variety and vibrant beauty.

Real transformation in our lives and in our world will occur as we translate these new visions and commitments, this "heart knowledge" into tangible practices within our own lives and communities. A colleague tells of her experience at a prestigious Ivy League school where she had gone to pursue graduate studies in the Buddhist-Christian dialogue. Committed deeply to the integration of intellectual and spiritual values, she met her advisor with great anticipation and asked him earnestly, "Well, Professor, what kind of meditation do you practice?" He replied in a disapproving tone of voice, "Oh, I am a scholar, not a meditator." Within weeks she had changed advisors, changed programs, and eventually even changed schools! If we are to take seriously the challenge of transformation, we must find ways to embody these new values in such a way that scholars can be meditators, theologians can be farmers, the young and the old can dig in the soil and walk barefoot in the dew, and together we can undertake the holy work of peacemaking and justice seeking. We can no longer afford to live as if our public and private lives, our spiritual and professional vocations, were disconnected. Our own healing and that of Earth's requires no less. Each of us can be a passionate lover of life, reveling in the breathtaking wonder of the moment and reaching new depths of meaning as we discover our true place within the web of life. The same love that causes our hearts to soar at the sight of a newborn baby or the vibrant colors of a New England autumn also can inspire our relationships, guide our meditative and recreational choices, and fuel our work for justice, peace, and the elimination of all forms of domination.

Even as we seek to embody a new way of living, with reverence for life and a renewed awareness of the sacred, we will necessarily face the reality that life involves suffering. In all its splendor and beauty, life is also about pain, decay, and death. In a culture that remains fixated on denying those realities through establishing an intricate pattern of control and domination, it may seem too frightening to face the reality of impermanence. Yet the human capacity to face the brevity of life and make meaning of suffering is again and again reflected in rich spiritual traditions that honor life's passings and celebrate life's turnings. Plumbing the depths of our own fears may be somewhat akin to a slow spring thaw, wherein new life gradually emerges out of places that previously appeared only barren and lifeless. There can be real freedom in facing our own finitude and releasing ourselves into the great wheel of life, which even death becomes but a passage, a transformation into new life. Perhaps, as suggested by TwoBears, Stephanie Kaza, Catherine Keller, and others, the starting point for an authentic, vibrant, life-sustaining spirituality is letting go of fear and immersing oneself in life's deepest mysteries. When we can love all of life's cycles, then we will join with joyful abandon the dance on the cosmic compost heap, one more holy creature the universe has graciously spawned.

With hearts that are learning to love again, we now stand at the cross-roads. As we look ahead to the path that is just beyond our seeing, we are reminded that to take "the road less traveled" will make all the difference. And deep within us we have a growing awareness of what we must do. It will not be an easy journey, but we will not be left to wander aimlessly or without hope. We will be nourished by the sheer mystery, vitality, and creative eruption that fundamentally characterizes life in all its varied forms. We will learn from Earth how to live in harmony and interdependence, and we will be sustained by a community of others who share our commitment to healing and justice. Along the way our greatest gift and deepest joy may be found in the "experience of being grasped by the mystery, beauty, and inherent value of life" and coming face to face with the divine presence that brings substance and form out of the swirling, turbulent energies of the universe and of our lives.

Through it all we will continue to reclaim our common heritage, our shared sacred story. It will be told through an array of myth and symbol, ritual and practice. This diversity of expression will speak to us of life's deepest mysteries and reveal to us the truths that connect us all. The priest and the aboriginal will both be its teller. The parent and the shaman, the ecofeminist and the native healer, the Buddhist and the scientist—all will give voice and substance, all will weave the new pattern that may serve as grounding for generations to come. Together they will offer us the place from which to do the holy work we must undertake to heal ourselves and our world. The necessity of such divergent voices harks back to earth's wisdom; that true health and lasting wholeness can be attained only in the midst of rich diversity. Then we will come to know, at the very center of our beings, the ancient and timeless wisdom the natural world continues to offer: "greater diversity leads to increased ability." And variety will be understood not only as the spice of life but indeed its very essence.

In the introductory pages of this book there arises a timely invitation "to get our spiritual bearings" as we seek to navigate through these unknown and perilous times in order to realize the longed-for Promised Land. The emerging truth of our time is that we have already arrived: we are living in the Promised Land; we are walking on holy ground. We need only take off our shoes, lift up our eyes, and reach out our hands to share the milk and honey and go out to wonder beneath the stars and know we are home.

Contributors

THOMAS BERRY —author of *Dream of the Earth* and co-author, with physicist Brian Swimme, *The Universe Story*—is a cultural historian, ecological ethicist, and self-styled "geologian." He is also a Catholic Passionist priest, a well-known speaker and scholar, and teller of the "new story" of who we are and where we are going. We are, he says, at the end of the Cenozoic and at the dawn of the Ecozoic eras, coming to understand what it means to be part of a universe that is alive and continuing to evolve.

PAUL BROCKELMAN is professor of Philosophy, distinguished university professor of Religious Studies, and director of the Religious Studies Program at the University of New Hampshire. His professional interests lie in phenomenology, particularly the phenomenology of religion. Author of a number of books, his most recent is an exploration of religious understanding and truth entitled *The Inside Story: A Narrative Approach to Religious Understanding*.

JOHN E. CARROLL is professor of Environmental Conservation at the University of New Hampshire. He holds the Ph.D. in Resource Development from Michigan State University and has served as a university professor in environmental studies for over twenty-five years, specializing in ecological ethics and values, ecology and religion, and international environmental policy and diplomacy. He is the author of *Environmental Diplomacy* and *International Environmental Diplomacy* and recently has edited, with Albert LaChance, *Embracing Earth: Catholic Approaches to Ecology*.

CALVIN B. DEWITT teaches courses in environmental science, wetland ecology and land resources. He serves as director of the Au Sable Institute in Michigan, where seventy students and faculty from U.S. and Canadian Christian colleges gather for an annual summer forum on ecology, environmental studies, and ethics. He has been a presenter and consultant at many international gatherings.

ALBERT FRITSCH, S.J., a Jesuit priest and environmental ethicist, is also an environmental scientist and director of Appalachia—Science in the Public Interest and Resource Assessment Services, both in Kentucky. He has authored numerous books, including *99 Ways to a Simple Lifestyle*, *Environmental Ethics*, *Down to Earth Spirituality*, and *Earth Healing*.

EVERETT GENDLER has been the rabbi at Temple Emanuel of Merrimack Valley, Lowell, Massachusetts, since 1971. He has served as Jewish chaplain and instructor in Philosophy and Religious studies at Phillips Exeter Academy, Andover, Massachusetts, since 1976. He holds a B.A. from the University of Chicago, and M.H.L. and ordination as Rabbi from Jewish Theological Seminary, and Doctor of Divinity degree, Honoris Causa, from the Jewish Theological Seminary.

STEPHANIE KAZA, assistant professor of Environmental Studies, University of Vermont, teaches environmental ethics, ecofeminism, and nature writing. She holds a Ph.D. in Biology from the University of California, Santa Cruz, an M.A. in Education from Stanford University, and an M.Div. from Starr King School for the Ministry. She is a long-time student of Buddhism and serves as chair of the board of directors for the Buddhist Peace Fellowship.

CATHERINE KELLER is associate professor of Constructive Theology at the Theological School, Drew University, Madison, New Jersey. She holds the Ph.D. in Philosophy of Religion and Theology from Claremont Graduate School and is author of *Apocalypse Now and Then: A Feminist Approach to the End of the World.*

ALBERT LACHANCE is a professional counselor and founding director of the Greenspirit Institute in Manchester, New Hampshire. He is the author of *Greenspirit: Twelve Steps in Ecological Spirituality* and co-editor (with John E. Carroll) of *Embracing Earth: Catholic Approaches to Ecology.* He is currently a doctoral candidate at the Union Institute.

JAY MCDANIEL is associate professor of Religion at Hendrix College, Conway, Arkansas. Having earned a Ph.D. in Philosophy of Religion and Theology from Claremont College in California, he joined the Hendrix faculty in 1979, where he serves as director of the Marshall T. Steel Center for the Study of Philosophy and Religion. He is a frequent author on topics relating to theology and ecology.

STEVEN ROCKEFELLER is professor of Religion at Middlebury College in Vermont, where he formerly served as dean of the college. He is the author of a number of books in philosophy and religion, including the recent and celebrated *Spirit and Nature: Why the Environment is a Religious Issue,* co-edited with John Elder.

TWOBEARS is a native of Vermont and a spiritual elder of the Abenaki people. He is one of a very few remaining speakers of the Abenaki language and a spokesperson both for the preservation of Abenaki culture and for native American spirituality.

TIMOTHY WEISKEL is currently the associate director of the Pacific Basin Research Center, Center for Science and International Affairs at the Harvard Kennedy School of Government. He is also the director of the newly created Harvard Seminar on Environmental Values at Harvard Divinity School. His most recent publication is *Environmental Decline and Public Policy: Pattern, Trend and Prospect.*

MARY E. WESTFALL has been chaplain to the University of New Hampshire and since 1990 executive director of the United Campus Ministry, an ecumenical Christian Ministry. She is an ordained Presbyterian minister, having earned an M.Div. from San Francisco Theological Seminary. Currently she is completing a Ph.D. in Environmental Ethics and Spiritual Values at the University of New Hampshire. In addition to local involvements, she has served on several national committees that deal with environmental concerns. She is a frequent presenter and workshop leader on topics related to ecology, theology, ecofeminism, and contemporary spirituality.

UNIVERSITY PRESS OF NEW ENGLAND publishes books under its own imprint and is the publisher for Brandeis University Press, Dartmouth College, Middlebury College Press, University of New Hampshire, Tufts University, Wesleyan University Press, and Salzburg Seminar.

LIBRARY OF CONGRESS CATALOGING-IN-PUBLISCATION DATA

The greening of faith : God, the environment, and the good life /
    edited by John E. Carroll, Paul Brockelman, Mary Westfall ; foreword
    by Bill McKibben.
        p.   cm.
    ISBN 0-87451-776-1 (cloth : alk. paper). — ISBN 0-87451-777-X
    (pbk. : alk. paper)
    1. Nature—Religious aspects. 2. Human ecology—Religious
    aspects. I. Carroll, John E., 1945- . II. Brockelman, Paul T.
    III. Westfall, Mary, 1960- .
    BL65.N35G47   1997
    291.1'78362—dc20   96-22547